INDUCED FISH BREEDING

INDUCED FISH BREEDING

A Practical Guide for Hatcheries

NIHAR RANJAN
CHATTOPADHYAY

Professor, Department of Aquaculture,
Faculty of Fishery Sciences,
West Bengal University of Animal & Fishery Sciences,
Kolkata, West Bengal, India

AMSTERDAM • BOSTON • HEIDELBERG • LONDON • NEW YORK • OXFORD
PARIS • SAN DIEGO • SAN FRANCISCO • SINGAPORE • SYDNEY • TOKYO
Academic Press is an imprint of Elsevier

Academic Press is an imprint of Elsevier
125 London Wall, London EC2Y 5AS, United Kingdom
525 B Street, Suite 1800, San Diego, CA 92101-4495, United States
50 Hampshire Street, 5th Floor, Cambridge, MA 02139, United States
The Boulevard, Langford Lane, Kidlington, Oxford OX5 1GB, United Kingdom

Notices
Knowledge and best practice in this field are constantly changing. As new research and experience broaden our
understanding, changes in research methods, professional practices, or medical treatment may become necessary.

Practitioners and researchers must always rely on their own experience and knowledge in evaluating and using any
information, methods, compounds, or experiments described herein. In using such information or methods they should be
mindful of their own safety and the safety of others, including parties for whom they have a professional responsibility.

To the fullest extent of the law, neither the Publisher nor the authors, contributors, or editors, assume any liability for any
injury and/or damage to persons or property as a matter of products liability, negligence or otherwise, or from any use or
operation of any methods, products, instructions, or ideas contained in the material herein.

British Library Cataloguing-in-Publication Data
A catalogue record for this book is available from the British Library

Library of Congress Cataloging-in-Publication Data
A catalog record for this book is available from the Library of Congress

ISBN: 978-0-12-801774-6

For Information on all Academic Press publications
visit our website at https://www.elsevier.com

Working together
to grow libraries in
developing countries

www.elsevier.com • www.bookaid.org

Publisher: Nikki Levy
Acquisition Editor: Patricia Osborn
Editorial Project Manager: Karen R. Miller
Production Project Manager: Lisa Jones
Designer: Victoria Pearson

Typeset by MPS Limited, Chennai, India

Dedicated to our loving and living GOD LORD JAGANNATH

CONTENTS

Part I INDUCED BREEDING—A SCIENTIFIC APPROACH TOWARDS MODERN FISH BREEDING PROCEDURE

PREFACE

Since joining the Department of Aquaculture, Faculty of Fishery Sciences, West Bengal University of Animal and Fishery Sciences, Kolkata, as Reader and subsequently Professor of Aquaculture, I have thought to review the status of the hatchery sector, as this the primary input that is fish seed is produced in the hatchery sector. I started my work involving the hatcheries of district 24 pargonas (pgs) north, WB, India, as primary activities relating to seed production in captivity started in this district. Dissemination of induced breeding technology was done in the said district through the establishment of one pioneer hatchery in the country. The initial establishment of hatcheries in freshwater sector was made with the help of a government official of the state hatchery in Naihati of the said district and state. During study I noticed that, except for one or two hatcheries, most of the hatcheries were established afterwards without any institutional or government support. Realizing the short-term profit, people from diverse sectors started venturing into the hatchery business by learning the technology from neighboring ignorant fish breeders, who are not in any way aware of the basic principles of the technology. This way mushroom hatchers came up from the late 1970s and are still appearing.

When we extended our study to two neighboring states (i.e., Assam and Bihar), we were astonished to notice that in one state (Assam) the farmers had learnt the technology by visiting different hatcheries of Bengal. In Bihar, the entire breeding operation in hatcheries was conducted by hired ignorant fish breeders from Bengal, particularly from the said district where primary work relating to breeding started in the late 1960s. Though there are some basic differences in the dissemination and establishment of hatcheries of the three leading seed-producing states in India, an overall negative approach was being developed among fish breeders of three states. Induced fish breeding, that is production of fish seed in captivity, was developed with the idea of production of the desired quality seed in captivity in times of need. The initial defect in the transfer of technology, without any institutional support, directs such a novel technology in a direction where quality and standards, the basic principles of the technology, are a matter of great concern. To date no institutional approach for a training program has been conducted to make these people apprised of the negative consequences developed in the sector due to their unscientific and illegal profit-making approaches. This trend attracts people from different sectors in the seed production business, mainly on the understanding of huge profits within a very short period of time.

Hence the people involved in the hatchery sector, being both illiterate and ignorant, have inadvertently used the technology since its inception only for short-term profit-making. In this endeavor they started indiscriminate hybridization by maintain a small number of the founder population. Rarely do they exchange broods and pedigrees, both of which stand as primary criteria for quality seed production. The potency of the gland is never considered as an important criterion for quality seed. This type of unscientific activities, by maintaining a small number of founder population, invited a great deal of negative genetic consequences such as inbreeding and depression, genetic introgression, genetic drift, and ultimately homogenization of the wild gene pool.

Due to the want of any certification procedure, the seed produced in the said hatchery sectors were not only qualitatively inferior but were affected with various congenital abnormalities and diseases. Thus the effective hybrid and inferior seed are transported not only to diverse geographical territories of the country but are also used to regenerate the wild stock through ranching programs. Practices like the entry of cultured fish into the wild are expected to affect the wild gene pool as fish reproductive isolation is not as strong as in other vertebrates. This has been elucidated through the capture of hybrids and deformed fish from different stretches of river. This practice to rejuvenate the wild population, as it is understood now, has already produced an overall negative impact through contamination of the native gene pool. While the wild population is declining alarmingly in the major rivers of India, this type of unscientific approach resulting in the entry of cultured fish, without any scrutiny, into the wild may pose a threat to very existence of the prized fish (i.e., rohu, catla, mrigal).

Intensive study for a period of 4–5 years covering three leading seed producing states of India are comprised into thirteen chapters. Fish breeding in captivity, which started initially by using prevailing congenial environment-like undulating terrain, a vast catchment area and bundhs in some regions, known as bundh breeding. Afterwards by the hormone induction application, from the discovery of induced breeding in 1950, that were extensively in use for production of quality seeds in captivity. Starting with traditional bundh breeding, and the intensive field study that was carried out to take note of the different breeding practices, including the scientific induced breeding technology. Special emphasis was given for collecting data on the history, modus operandi, status of the technology, and its modification from the time of its implementation as several innovative approaches on the part of fish breeders. In India, since the discovery of induced breeding technology and its subsequent dissemination to field, without maintaining any code of practice of transfer technology, has turned such an epoch making technology to the fresh water aquaculture sector. The fish farmers and a large number of illiterate people from other occupations, realizing the large short term profits, started practicing the technology for profit-making propositions, by learning the technology from neighbouring illiterate fish breeders.

This sort of unscientific transfer was not restricted to one state instead the people from different states visited the pioneer state like West Bengal and carried away the technology without any scientific back up. This resulted in gross misappropriation of technology for a period of 50–60 years, and has created a lot of negative genetic consequences like inbreeding, genetic drift, genetic introgression, and ultimately resulted in contamination in the wild gene pool. The author has, time and again, tried to elucidate such an irreparable, damaging, and unscientific misappropriation on the part of such a novel technology, solely for profit. Now, though such a pernicious activity on the part of fish breeders are known to scientists, policy makers, and government officials, however, nothing has been seriously initiated yet. Although there are no paucity of scientific publications at national and international level.

It is interesting to note that fish breeders are innovative and in times of difficulty, have developed innovative technologies to overcome situations. One such example is the discovery of suitable a chemical substance to remove the glue from eggs of fish that lay adhesive eggs. Recently, a new approach in the name of conservation hatchery is in use to replenish the loss and rejuvenation of wild stock. I appreciate the scientific endeavour from different positions, regions and states to find an immediate approach to save the prized fish of India.

INTRODUCTION: PREEXISTING (TRADITIONAL) AND MODERN FISH BREEDING METHODS IN PRACTICE AMONG FISH FARMERS

Aquaculture in India and neighboring countries, such as Bangladesh and Pakistan, mainly constitutes Indian major carp (IMC), namely, the catla (*Catla catla*, Hamilton), the rohu (*Labeo rohita*, Hamilton), the mrigal (*Cirrhinus mrigala*, Hamilton), and rarely Kalbasu (*Labeo calbasu*, Hamilton) as this is being threatened. These carp contribute approximately 75% of the aquaculture production in India (FAO, 1997). As these carp are economically very important, research on cultivating these species of fish was initiated in India during the early 1950s to study and understand the biology of these economically important species, particularly the major carp, and to develop suitable technologies for various farming systems.

India possesses nearly 11% of the world's 20,000 known species of fish, and is one of the richest nations in the world with regard to genetic resources of fish, which are distributed over a network of perennial river systems. Considering the culture fisheries or aquaculture, fish culture in India is almost as old as in China; the average per hectare production per year remained as low as 0.6 tonnes until the 1960s. This was due to lack of proper technology, as the fish farmers followed traditional or empirical methods for farming. Before the introduction of fish breeding technology following the Chinese method (Linpe method), fish seeds were collected from the wild by collectors using a traditional scoot net. As this resulted in the collection of a mixture of both the economic and uneconomic species, this wild collection failed to create a significant impact on the culture sector, which still remained quackery.

In the backdrop of such a state, induced breeding, primarily of IMC, is a great landmark in the aquaculture development in India and also made blue revolution possible. After its development at the CIFA center of Bhubaneswar, Orissa, India, the technology was not disseminated properly to the farms following the principle of transfer of technology. The farmers of Bengal started using the technology without having primary training on fish genetics and basic principles of the technology. The farmers of Assam and Bihar learned the technology afterwards by visiting Bengali fish breeders. This indicates that the fish breeders of the three states started using the technology only for their profit and short-term gain, due to the paucity of space and money the fish breeders adopted various injudicious ways like mixed spawning, indiscriminate hybridization, and use of immature brooders with skewed sex ratio and also multiple breeding. As we know, mixed spawning leads to hybridization inadvertently as the sexual isolation of fish is not stringent. Hybrid fertility is observed in the case of intergeneric and reciprocal hybridization. Again it is well-known that inadvertent hybridization and backcrossing of F1 hybrids, which is the usual practice of the hatcheries of the three states, would cause genetic introgression resulting in contamination in the wild gene pool of these prized fish of India. Experience with hybridization of domestic common carp with its wild ancestors resulted in the contamination of both the stocks and in the deterioration of their economically important traits. Along with this, inbreeding is a common practice as the fish breeders maintain a small number of founder populations. It is known that one generation of inbreeding resulted in increased high deformity (37.6%) and decreased food conversion efficiency (15.6%) and fry survival (19%). Due to the basic defect in adoption of the technology, the potency of the gland was never considered as a criteria for fish-breeding purposes. Due to this, the actual number of the breeding population decreased successively. Now it appears that there is a need for an objective study to develop a breeding strategy and selective breeding program. The genetic consequences of mixed spawning should be assessed under experimental conditions and by surveying the natural populations. The problems of stock contamination may be assessed with subtle tools of isozyme and DNA polymorphism.

The primary goal of the seed production industry should be to produce "quality fish seed" and subsequent distribution of the same among the farmers for culture or further growing out. In scientific terms, "quality seed" may be defined as those having better food conversion efficiency, high growth rate potential, better ability for adapting to changing environmental conditions, disease resistant (Padhi and Mondal, 1999), and fetching a high market price. Seed production is a complex procedure and requires extensive knowledge of fish physiology, fish endocrinology, and also knowledge of the environmental influence on maturation, spawning, and hatching. It also requires knowledge of hormone-induced breeding (Atz and Pickford, 1959), population genetic principles, hybridization, and brood stock development, the art of hatchery and nursery management along with other relevant aspects needed to produce quality seed in captivity. Ideal seed production technology encompasses the following criteria:

1. Brood stock collection and management along with replacement of stock from nature at certain intervals;
2. Collection of good-quality (having the right potency) pituitary gland and its preservation in ideal conditions;
3. Selection of ideal breeders;
4. Maintenance of an ideal spawning and hatching environment;
5. Artificial breeding;
6. Hatchery and nursery management.

The practice of breeding is the practical aspect of a thorough understanding of the science of fish physiology, endocrinology, genetics, and the breeding environment in captivity. The fish breeders, being totally unaware of the scientific basis of the technology, started using the technology empirically only for profit and short-term gain for a fairly long time. Again the farm-raised spawn and fry are not only transported to diverse geographical territories but find their way into natural systems through ranching and natural phenomena. All these practices over the years, as mentioned, led to the development of a series of negative consequences through inbreeding, genetic drift, indiscriminate hybridization, mixed spawning, and genetic introgression. Again, the gene pool has been modified consciously by selective breeding of fish for some commercial gain such as increased growth rate or disease resistance ability. On the other hand, due to lack of awareness of the scientific aspects of fish breeding and fish genetics, the breeders invite such consequences in the gene pool of farm-raised fish, which in the course of time will contaminate the native gene pool. This is evident from the collection as well as availability of phenotypically different fish species of IMC (rohu, catla, and mrigal) both from farms and natural harvests.

Fish breeding in captivity through hormone injection or manipulation of available environmental conditions dates back to 1882. Before successful standardization and dissemination of induced breeding technology, there happens to be evolutionary history starting from bundh breeding to portable circular hatchery through a number of intermediate models. Just before the implementation of induced breeding technology 70% of the seed requirement was met from the fish seed raised out of bundh breeding. As the fish seed raised here was mainly through natural process and environmental maneuvering, the quality was good and there were no complaints from the farming community regarding poor growth, delayed maturity, deformities, and the many other unwanted consequences as are faced now after introduction of induced-breeding technology.

Linpe method of induced breeding: Collaborative efforts by Chinese and Canadian researchers led to the development of a technology that revolutionized fish seed production in captivity by injecting GnRHa in combination with domperidon or pimazide (a dopamine antagonist), which is known as Linpe method (Lin et al., 1988). It is established now that GnRH alone failed to induce fish due to the presence of endogenous dopamine. Dopamine occupies the GnRH receptor and thus blocked its action on pituitary gonadotroph cells. GnRH induces gonads while the drug domperidon inhibits the action of dopamine—a substance secreted by fish during ovulation. In comparison to pituitary extract (which possesses a poor shelf life), which is subjected to protein denaturation and infection by microorganisms, GnRH is more effective for successful spawning. Other negative aspects are the collection of glands from dead fish by ignorant gland collectors from fish markets along with unscientific preservation procedures. Again, pituitary induction requires the application of two doses at an interval of 6 h, indicating that it is more hazardous and time-consuming compared with GnRH induction. With traditional fish spawning methods, carp, e.g., are raised and killed to produce a pituitary extract used to induce spawning. Many fish are sacrificed in the process and the extract has a poor shelf life. The technique also requires that fish are injected at two separate intervals to induce ovulation.

The new method reduces the cost of production, increases the supply of seed fish, and is more convenient. Rates of spawning, fertilization, hatching, and survival were significantly higher in research trials than could be achieved with pituitary injections. The hormone and drug can be introduced together, which means that brood fish stocks are handled only once, reducing the risk of disease or damage to the fish. This method does not alter the reproductive cycle of the fish, and the fertility and viability of offspring are normal. The solution does not require refrigeration and has a long shelf life. It has been tested on a wide range of fresh, salt, and brackish water species, including carp, bream, salmon, catfish, loach, and others.

It is tempting to generalize about the superiority of the Linpe method for all cultured fish, but because comparative experiments under field conditions—GnRHa–domperidone versus GnRHa alone—have only been done for carp, so it is still too early to go with such inferences. Some researchers have tried to put fish into categories that reflect the strength of dopamine inhibition, ranging from cyprinids (strong dopamine effect) to salmonids (weak dopamine effect). The danger of making this kind of categorization relates to what we have already said about differences in the effect of GnRH itself. Maturation at the peak of its stage along with congenial environmental conditions can outweigh any advantage or disadvantage of a particular treatment.

The commonly voiced view that marine fish do not require domperidone along with GnRHa requires more proof; in milkfish and mullet, e.g., two of the most important warmwater marine species, there are no published reports of its having even been tried. Until use of the Linpe method is more widespread, we will avoid such lists. In species that become fully sexually mature in captivity and respond to GnRHa readily—many salmonids fall into this category—a dopamine antagonist is not needed. In other species—the best evidence is still from cyprinids—even though they will spawn with GnRHa alone, delay to ovulation is shorter and more predictable when domperidone is added. Administering the two drugs is easy, with a single injection of a mixture being as effective as two separate injections. This has led to the manufacture of a commercial GnRHa–domperidone spawning "kit" that combines the two in a single solution (Ovaprim). Enterprising fish farmers can of course always opt, as in Thailand, to buy GnRHa and domperidone as over-the-counter pharmaceuticals, and reconstitute them for injection into fish.

According to Dr. Lin Hao (Ran of China's Zhongshan University), the Linpe method has become "…more and more popular in Chinese fish farms and has replaced the traditional fish spawning methods in recent years." Dr. Lin has established a commercial operation to sell the active compound in China through the Ningbo Hormonal Products Factory in Ningbo City. Commercialization was identified as a specific objective in phase II of the research project. In addition, Syndel International Inc. has submitted an application (pending) to register Ovaprim at the regulatory agency in China, after running clinical trials in Wuxi, Beijing, and Harbin in 1994. Linpe method (domperidone/sGnRHa) of induced spawning of cultured freshwater fish is used in many countries, leading to commercialization of the method; to determine the effectiveness of sGnRHa and domperidone in induced ovulation and spawning of marine teleosts; to determine the effects of aging on reproductive function of key species in the Chinese freshwater polyculture system; to determine means of increasing growth rates of cultured fish; and to continue the training of young Chinese scientists in relevant disciplines.

Although there will never be a standard method for spawning all species, culturists working with a single species can standardize methods by systematically eliminating sources of variability and using the lowest effective dose. Effective doses of GnRHa and domperidone vary widely and are not comparable because of differences in species, temperature, state of maturity, and GnRHa. The trend is toward single injections and, although GnRHa doses between 1 and 100~g/kg have been effective, culturists should aim for the 5–20 ~g/kg range. Domperidone is usually effective at doses of 1–5 mg/kg. To facilitate economical use of GnRHa, without the need for tedious weighing of tiny amounts, it is best to buy preweighed small amounts of the hormone (e.g., 0.5 or 1 mg aliquots) and prepare a concentrated stock solution (e.g., 1 mg/mL) in sterile water in the original container. Appropriate amounts of a more dilute solution in 0.7% NaCl can then be prepared at the time of injection. GnRHa is most stable as a dry powder, but the sterile stock solution can also be kept for several months if frozen. Domperidone and pimozide are not readily soluble in water and are sensitive to oxidation. They are best used as a suspension in 0.7% NaCl containing 0.1% metabisulfate as antioxidant, or can be dissolved (and injected) in propylene glycol. Commercially available domperidone tablets for humans (Motilium 0) have been powdered, dissolved in propylene glycol, and used successfully in induced reproduction of fish (Fermin, 1991).

Bundh Breeding in Captivity

The correct information about the first establishment of fish seed production through bundh breeding techniques is not available as per the existing records among the fish breeders of Bankura District of West Bengal, India. The first report of bundh breeding of carp was in 1882 by a private pisciculturist, but the systematic practice of collection of eggs started commercially in 1902 at "Simlapal" village, presently Simlapal Development Block in Bankura District of West Bengal, India.

Out of the two bundh breeding practices, though systematic dry bundh breeding started in 1926, but due to erratic and profit-making approaches from the sector, this unique, natural and quality seed production industry became irresponsive. It was reported that sterility developed in the bundh and the seed production industry became irresponsive due to the absence of a congenial breeding environment. As the supply was markedly decreased so further scientific approaches to rejuvenate this novel seed production sector were initiated by a group of scientists by introducing "sympathetic breeding." The objective was to inject 10–15% of the broods and to release hormone-induced and noninduced broods together at the catchment area at the end of the bundhs. This resulted in spawning of both the induced and noninduced fish at the less deep region surrounding the central pond. Thus a new dimension was added to bundh breeding in the year 1960. The fish breeders, realizing the profitability and ease of the sympathetic breeding procedure, started implementing the procedure in bundhs and more than 1500 bundhs were involved in the production of fish seed by using a combination of both the natural and artificial methods of breeding. The fish seed produced through this new approach were qualitatively as good as before, and had contributed about 60% of the total seed production of the country. This continued until 1970 but efforts were on to further modify and improve bundh breeding by the addition of the Bangla bundh concept and also by developing a system of steady water flow by constructing a cement cistern resembling a community breeding pool. The artificial bundh is so constructed that the bottoms of the artificial bundhs possess a gentle gradual slope from the deep and to the shallow end. The flow within the Bangla bundh is maintained by installing an inlet at the deep end and an outlet at the shallower end. This indicates that from the very beginning, the fish breeders sometimes of their own accord and sometimes at the advice of scientists and government officials tried to overcome the impediments faced by this sector through implementation of innovative approaches.

What Is a Bundh?

The term "bundh" originated from the Bengali language and means protection of low-lying area by erecting dykes or embankments. The undulated nature of land in the districts of Midnapur and Bankura helps the formation of a particular type of tanks known as "bundhs." During rainfall, water from upland catchment areas rushes downwards and finds its way through a depressed land having two raised sides. The third side is an embankment to hold the water for agriculture and other purposes by the local people. This is how bundhs originated in those districts. In fact, a bundh is a shallow depression having a dyke on one side and a vast catchment area on the other sides, gradually sloping down into a depression. During rain the bundhs are flooded by the flushing of rainwater from the surrounding catchment areas. Alikunhi et al. (1960) stated that "bundhs" are specialized ponds where a riverine condition is simulated by constructing embankments against large catchment areas and subjected to rapid flooding during monsoon due to gravitation.

Area of Bundh

Different authors have advocated different proportions between the deeper zone (bundh proper) and the catchment area. In West Bengal, the ratio between the pond area (depression) and the catchment area is 1:5; in Madhya Pradesh it is 1:25, whereas in Gulati it was considered to be 1:10. The difference is due to the variance in annual rainfall. It has also been noticed that bundh breeding is being conducted in small ponds less than $14,000 \, \text{ft}^2$, with a shallow area having no catchment area; only a depressed area situated below the level of an irrigation canal, or water is lifted by pump and allowed to pass through this depressed embanked area, have successfully conducted bundh breeding operations in West Bengal.

Different Parts of a Typical Bundh

A typical bundh consist of the following parts:
1. The catchment area;
2. The depression or pond;
3. Moan;
4. Bullan;
5. Cherra.

The vast area from where rainwater rushes toward the pond is called the *"catchment area."* The shallower area adjacent to the pond opposite the catchment area where generally fish breed is called the *"moan"* and the passage through which water escapes outside the bundh is called the *"bullan."* The bullan is guarded by a split-bamboo screen, called a *"cherra,"* to prevent the escape of fertilized eggs.

Topography of a Bundh

Almost all bundh breeding operations are conducted in the laterite belt of West Bengal, Madhya Pradesh, and Bihar, where there is a slight variation of annual temperature as well as rains. The following factors of topography are considered as suitable to get successful results.

Types of Soil

The texture and chemical composition of soil have a great influence on the breeding of IMC. Laterite soils contain a lower percentage of organic constituents with red, yellow, or orange tint due to the presence of iron oxides in various degrees of hydration. Recent experimental work on bundh breeding provides a definitive role of soil for increasing breeding and hatching rates.

Impact of Regional Soil on Hydrobiological Parameters of Breeding and Hatching Bundh

To evaluate the role of regional soil at the soil–water interface, hydrobiological parameters of water, i.e., temperature, pH, transparency, dissolve oxygen, biological oxygen demand, hardness, alkalinity, and salinity of both the breeding and hatching bundh were noted by following a standard method (APHA) at an interval of 7 days, i.e., on day 0, day 7, day 14, and day 21 by adding different quantities of soil. Three sets of experiments were designed in three aluminum containers, numbered as 1, 2, and 3. After filling the containers with 20 L

Figure 1 Farmers engaged in mixing special soil with the water of a hatching bundh before releasing fertilized eggs to ensure maximum hatching.

of water 50, 100, and 150 g of soil were added successively to containers 1, 2, and 3. A control container was maintained in which no soil was added. Each container was stocked with five pieces of *Tilapia* fingerlings. From day 0 onwards water samples were collected from the four tested containers and the above parameters were tested in a laboratory. The resultant data were compared with controls to evaluate the role of soil in increasing fertility and hatchability of carp eggs by comparing the data (Banerjea, 1967).

The soil when mixed with water in the hatching pond offers extra buoyancy to the fertilized eggs, maybe by increasing the density of water, and the eggs remained floating for a longer period. This enhances the hatching rate, otherwise the fertilized eggs settle at the bottom and hatching is impaired. Besides this, soil removes glue from the adhesive eggs (Fig. 1).

Slope

The slope gradient of the catchment area is ideal for a bundh. The bottom contour of the breeding ground should also provide a gradient slope toward the outlet of the bundh. The catchment slope has a significant importance in many ways: firstly, in computing the time required for filling up of dry bundh; secondly, in studying the soil water contact angle; thirdly in calculating the degree of erosive power of rainfall; and finally, in calculating the run-off coefficient (Fig. 2).

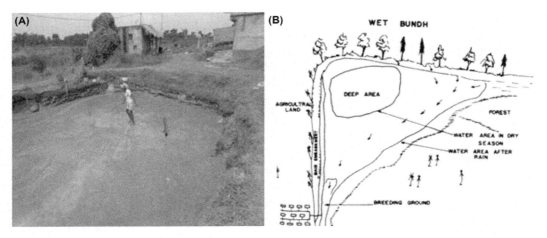

Figure 2 (A, B) Wet bundh located in undulating terrain of Bankura districts.

Catchment Area

The catchment area may be a forest or agricultural land with gradual sloping toward the deeper zone (bundh proper). It appears that the area of catchment mainly depends on the intensity of annual rainfall; where annual rainfall is less, more catchment areas are required to make the overflow and current continuously in the bundh proper.

Bundh Proper (Deepest Area)

In the wet type of bundh, the depth should be more than 6 ft and generally the deepest area should be more than one hectare in area. A stock of fish is maintained in that area round the year. In the case of a dry bundh, the bundh proper varies according to the quantity and size of brood fish to be maintained per operation. In general, on an average 10 katha ($700\,m^2$) at top level and $7^1/_2$ katha ($500\,m^2$) at surface level is maintained. An inlet having a diameter of 30–40 cm with a discharge capacity on an average of 500 L ($0.5\,m^3$) per second is installed at a height of 60 cm (from ground level) and an outlet at a height of 15–20 cm (from ground level) is fixed at the opposite end. But in Bankura and Midnapore (Garbeta) instead of setting hume pipes, people construct an inlet by cutting a portion of the embankment and the inlet is guarded by a split-bamboo screen. As a precautionary measure, an emergency outlet at the upper ridges of the dry/wet bundh is made to let out excessive water in case of sudden unmanageable rush of water (flooding).

Size and Height of Embankment

The engineering aspect includes proper designing of the bundh, which includes the laying of a bottom contour of breeding grounds as well as the height and width of the embankment, fixation of inlet and outlet, etc. Generally, the embankments are constructed with a slope gradient of 1:4 and a height of 4 ft from ground level.

Breeding Environment

Different authors have emphasized different factors, which are sometimes contradictory for positive results in spawning in bundhs. Physical factors like water current, raising of the water level, rainfall, flood, depth of water in spawning ground, laterite soil and its color, turbidity, temperature of water, storm, thunders, etc. have been regarded by many authors as factors that influence spawning of fish in bundhs. Some authors are in favor of considering chemical factors like oxygen (4.2–6.8 ppm), pH (7.2–8.2), total alkalinity (below 2 ppm), and free CO_2 (2 ppm) as responsible for breeding of fish. A few others still consider the silt particles found on the body of spawners, while some are in favor of considering a positive role of minerals in solution or in suspension in water, as well as the electromagnetic properties of water.

Types of Bundh

Bundhs are nothing but a specialized type of pond where riverine conditions are simulated. They are constructed in the middle of a vast low-lying area, with proper embankments and receive large quantities of rainwater after heavy rains. Bundhs are provided with an outlet for the overflow of excess water, and shallower areas which serve as spawning grounds for the fish. A large number of similar kinds of bundh-type tanks are found in West Bengal and Bihar. During recent years, bundh breeding has been conducted on a large scale in Madhya Pradesh. Bundhs are generally of two types:
1. Perennial bundh also called "wet bundhs."
2. Seasonal bundhs or "dry bundhs."

Wet Bundh

This is a perennial pond situated at the slope of a vast catchment area, with an embankment on three sides. The main pond retains water throughout the year, but its shallow marginal areas dry up during the summer months. The bundh has an inlet toward the high catchment area, and an outlet at the opposite lower side. The bundh is usually flooded with water from the upland area after heavy showers. The shallow area of the bundh, called the "moan," is inundated and excess water flows out. The outlet is protected by bamboo fencing. The breeders, which are either grown in the perennial pond or released from other ponds, are stimulated by the flow of silt-laden and well-oxygenated rainwater and subsequently spawn in the shallow areas (Fig. 2).

Dry Bundh

A dry bundh is a seasonal shallow pond enclosed by an earthen wall (embankment) on three sides. During the monsoon season, rainwater rushes from the vast catchment area and accumulates in the pond. Breeders from nearby ponds are introduced into the shallow ponds. Breeding takes place when, after a heavy shower, the bundh is flooded with fresh rainwater. It has been observed that the carp migrate to shallow water, when, after a little sexual play, they spawn. The eggs are collected by means of a mosquito net and transferred to small ditches or cloth hapa of the size of 4' × 3' × 3' for hatching. After 3–5 days, fry are transferred to nurseries (Fig. 3A and B).

After successful breeding of carp in the dry bundh in *Sonar Talliya* in Madhya Pradesh (India), several dry bundhs of improved designs were constructed. The modern bundhs are permanent (locally called pacca) masonry structures and are provided with a sluice gate in the deepest part to facilitate dewatering the bundh after each spawning. This enables easy removal of spent fish and collection of fry. There is storage for stocking breeders and a set of cement cisterns serving as hatcheries. In some cases an observations-cum-shelter tower is also constructed near the bundh (Fig. 4).

Figure 3 (A, B) Collection of eggs from a breeding bundh by net.

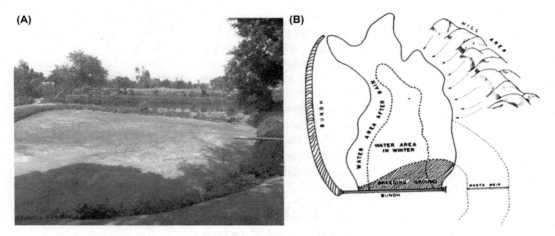

Figure 4 Dry bundh: (A) a photo indicating the slope gradient from the catchment area and (B) drawing of a dry bundh.

Spawning in both types of bundhs occurs after continuous heavy showers, when a large quantity of rainwater rushes into the bundh. As soon as water accumulates in the dry bundhs, a selected number of breeders, in the ratio of one female to two males (1:1 by weight) are introduced and a constant vigil maintained from the observation tower. Spawning may occur in the evening, at night, or in the early morning depending on the availability of congenial environmental conditions. According to Alikunhi et al. (1964), mrigala and rohu start breeding in the morning, while catla spawn from about noon to evening. Mrigala and rohu spawn in marginal shallow areas, while catla remain confined to relatively deeper water due to their deep body. Males and females both exhibit sex play for a short time during which their bodies are twisted round each other. The pressure on the abdomen results in the exudation of ova and milt. Eggs are laid at intervals during which time the pair keeps moving close to each other. After spawning, the spent fish move to the deeper water. Eggs are collected with mosquito nets and transferred to hatching hapas or cement cisterns. According to Alikunhi et al. (1964), 50–68% of the fertilized eggs are recovered from dry bundhs in Madhya Pradesh, and survival in hatching hapas in 30–35%, but this rate has been improved to 97% by using cement cisterns as hatcheries.

Although Indians carp are induced to breed in the special type of ponds described above, there is no information regarding breeding of Chinese carp in bundhs. It is believed that the Chinese carp, which have more or less similar breeding habits to the Indian carp, can also be bred in bundh-type tanks.

Several factors are reported to be responsible for spawning of major carp in bundh-type tanks. According to Hora (1945) and

Khanna (1958), spawning is stimulated by heavy monsoons that flood the shallow spawning areas. It appears that the availability of shallow spawning grounds inundated with fresh rainwater is an important factor in stimulating the breeders. According to some workers, spawning occurs in still waters, while others believe that strong currents on the spawning grounds favor breeding. The favorable temperature of water for spawning has been found to vary between 24°C and 32°C under various environmental conditions. Generally, cloudy days followed by a thunderstorm and rain, are regarded to influence spawning. Other factors like pH, alkalinity, and high oxygen content may be of secondary importance, and are associated with floods.

Operation Technique of a Dry Bundh for Production of Spawn

Brood fishes are introduced into bundh from a stocking tank with a sex ratio of 1:1 (male:female) according to the size of the bundhs. Before releasing into the bundhs, the maturity condition of both the sexes is properly and carefully examined. The weight ratios between both the sexes are maintained as 1:1 as far as is practicable. After introduction of brood fish, water is allowed to enter into the bundh through an inlet and to escape through an outlet which is guarded by a split-bamboo screen and hereby a current is produced inside the bundh area. When the first sign of coiling, i.e., encircling of a female a by male is noticed, the entry and escapement passages is blocked. Subsequently, spawning takes place. Generally the breeding operation is undertaken on an evening/night as the congenial breeding temperature and environment exist at that time.

On the next morning, first of all spent fish are removed by dragging a larger mesh-sized drag net. Then the fertilized eggs are collected by "chat jal" or mosquito nets. It is then placed for hatching in a series of earthen pits, called "chobba or hapa," having the average size of $2' \times 1' \times 1'$. A narrow nullah (drain) runs in between two series and periodically water is allowed to enter from the nullah into each chobba. The quantity of eggs placed in each chobba depends upon the percentage of fertilization, i.e., a higher percentage of higher quantity in each chobba; generally, a half to 1 L of eggs are placed into each chobba.

After hatching out (which depends on water temperature) hatchlings are collected from each chobba by a gamla/silk cloth and transferred to bigger-sized earthen pits called "hammar." This is generally double the size of the chobba. Likewise, a narrow drain runs in between the series of hammars and periodically water is supplied to each hammar from that drain. After 3 days, the spawn are disposed of from the hammar.

Improved Methods for Bundh Breeding

The method of bundh breeding has been improved (Chattopadhyay et al., 2013) by administration of pituitary gland extract to ensure the positive results of full spawning of each female spawn (Sundararaj et al., 1972a). Out of the whole stock 10–15% of the brood fish (by number) are injected. The male and females are injected at 1.5–2 and 4–6 mg/kg/body wt., respectively, as a single resolving dose or sometimes in the case of females the total dose is administered in two equal installments. The current inside the bundh is maintained by a pump and fertilized eggs are placed for hatching into double hatching hapa, consisting of one inner hapa within an outer hapa.

Spawning

Spawning usually takes place about 6–9 h after the injection, when a single high dose is administered. In the case of a single dose, evening is the most convenient time for injection. Though the spawning is expected around midnight or a little later (if the injection is given at the evening), the hapa should not be disturbed until the next morning for observation of eggs. Normally if there is no spawning even after the knockout dose, a second injection is not recommended. But if the condition of the breeders is found to be encouraging, a second injection at the rate of 4–6 mg/kg body wt. of the female and 2–3 mg in the case of the male may be given in the morning. Spawning is usually expected in another 3–6 h.

In the case of two injections (preliminary and a second dose to the female and only one dose to the males), the preliminary injection is usually given at noon, and the second 6 h later, in the evening. Though the fish is usually expected to spawn within 3–6 h of the second injection (i.e., at midnight if the second injection is given in the evening), the hapa, in this case also, is kept undisturbed till the next morning. If there is no response even after the second injection, a third injection at the rate of 4–7 mg/kg body wt. to the female and 2–3 mg/kg body wt. to the male may be administered in the morning, depending on the condition of the breeders, ecological conditions, and the local environment at the hatchery. Spawning may be expected in another 2–5 h. The third injection should usually be avoided, as repeated handling may cause injury to the breeders. In almost all cases, the injected fish become excited and active within 2–5 h of the first injection, when they start jumping and become restless.

There is no hard-and-fast rule regarding the time of the injection. The injection may be given at any time in the day or night, because spawning may occur at any time during the day or night. But since a low temperature is helpful for spawning, it is advisable to push injection on a cool day or in the evening or night when the temperature is fairly low.

Modern Technique for Bundh Breeding in Nonlaterite Belt

The introduction of induced breeding by hormone injection also opened a new avenue for breeding of fishes in small tanks of nonlaterite soil areas of some states. The fish are injected and liberated in such tanks and water is rushed in to create a circulation or current inside the tank and the fish are found to breed nicely. The same thing was developed subsequently by introduction of "Bangla bundh." A long elongated shallow cemented cistern-like tank having a dimension of $75' \times (20'-25') \times (2'-4.5')$ approximately, the bottom of which is usually lined with a few inches of sand at the time of each breeding program. There is also a barm separating the deeper zone of the tank with the shallower zone where the fishes usually breed, after injection and rush of water.

1. Inducing agent

The fish breeders of Panchmura (Bankura district, India) are now using WOVA-FH (chemically salmon gonadotrophin-releasing hormone analogue and domperidone), a synthetic inducing agent (Sundararaj et al., 1972b). Just before administration of injection, and 4 h after their collection, the brooders (both male and female) are caught from the brooders' pond and kept in a net or hapa to be easily taken for injection (Fig. 5). The fish breeders

Figure 5 Injection of WOVA-FH to a catla fish on the bank of a breeding bundh in Panchmura.

exploit sympathetic breeding by injecting only 90 brood fish out of 300 (30%) considered for a single game. For this the fish breeders prepare 90 mL of diluted solution by adding 80 mL of saline water to 10 mL of inducing agent (WOVA-FH). Now, out of 300 broods (150 male + 150 female) the fish breeders inject the diluted inducing agent into only 90 pieces, i.e., only 30% of total brood fish are induced (30 males + 60 females). Both the males and females are injected with only one dose at the rate of 1 mL diluted solution per kg body weight in the case of IMC. However, the percentage of brood fish to be injected and the dose of inducing agent varies depend upon temperature, maturity of brood fish, and the period of breeding season (i.e., premonsoon, monsoon, and postmonsoon).

2. Preparation of breeding bundh

A bundh is a shallow pond (length 50 ft × width 30 ft × depth 6 ft) having a slope from one end to the other (Figs. 1 and 2) in such that in the upper end the water depth is 4 ft, while at the lower end it is 6 ft and having an embankment. Generally the breeding ground is sandy. Water is supplied to the bundh from a shallow pump or from the nearby rivulet by a diesel pump or by an electric motor pump if available. The water height is maintained at 2.5 ft at the upper end and 3.5 ft at the lower end. Along with the inflowing water toward the bund, the farmer used to mix one type of special soil, available in this area and collected from 10 ft below the surface soil. They are convinced that this soil imparts buoyancy to the fertilized eggs to float in the water after spawning. According to fish breeders, this also increases the hatching percentage in the bundh (Fig. 6). The breeding bundh has the capacity to hold about 200–300 kg brood fish for a single game. However, the number and weight of brood fish released to the breeding pond depend upon the availability of brood fish and demand of fish seed. After completion of one game, the entire water is drained out and the breeding ground is washed thoroughly and left for 1 or 2 days for removal of bad odor and is again prepared for the next cycle.

3. Breeding, spawning, and fertilization

After injection the brooders are released again into the breeding bundh and within half an hour, the injected male and female exhibit breeding behavior, which includes sporting. Rapid underwater movement followed by splashing of water is indicative of the sporting mood of the breeders. Sporting continues for 2–3 h in the shallow region of the breeding bundh having a depth of 1 ft. Breeding ends in spawning, the females start releasing eggs at the shallow region and the males being enticed start releasing milt over the eggs, which results in fertilization. Sometimes spawning

Figure 6 The characteristic soil available in the area of bundh breeding plays a significant role in enhancing fertilization, hatching, and is also used for removing adhesive glue from the egg.

continues for 8h and both the sporting, spawning, and fertilization depend on the temperature of the air and water. Even after completion of spawning, fish breeders allow the brooders in the breeding bundh along with fertilized eggs for 3–4h and kept a strict vigil on the overall conditions of the pond. After some time the breeders start stirring the water by moving within the pond to declamp the egg mass. The fish breeders conduct this stirring movement around midnight of the first day or day 1.

From the second day onwards, as the development of the eggs proceeds, a dot appears in its middle and after 2–3h, the dot changes to a Bengali 5-like structure. Along with this the embryo indicates rapid movement within the egg shell. Immediately the fish breeders start transferring brood fish into the stocking pond and release the developing eggs into the hatching bundh after collecting the eggs by net (Fig. 4A and B). The farmers of Panchmura area transfer the eggs from the breeding bundh to the hatching bundh by cycle messenger (Fig. 7) with aluminum handi (Fig. 8).

4. Hatching

Hatching is a process by which spawn or hatchlings are released from the fertilized egg. For hatching, farmers of Panchmura area follow the following process to get maximum results.

5. Preparation of hatching bundh

Hatching bundh is nothing but a series of muddy or earthen pits of various sizes, such as 38 × 15 × 4ft, 23 × 15 × 4ft, 12 × 7 × 4ft. The bottom of the hatching bundh is convex in shape, and so

Figure 7 Egg transfer to the hatching bundh by cycle carrier.

Figure 8 Loading of fertilized eggs in handi for transfer to nursery.

holds more water in the middle than its surroundings. The water depth of the hatching bundh is 2.5 ft in the middle and 1 ft in the surrounding. The water which is used for the hatching process is from shallow or nearby canals or rivules. Before releasing eggs into the hatching bundh, fish breeders allow the bundh to be sundried for 3–4 days and fill it with water up to the desired level as required

Figure 9 Outlet for draining excess water.

for hatching. The fish breeders avoid use of any chemicals, cleaning agents, or fertilizers (Fig. 9).

6. Hatching practice

After preparing a hatching bundh farmers release the eggs to the bundh. The amount of eggs released for hatching depends on the size of the bundh and the environmental conditions. Fish breeders measure the eggs by using a special aluminum handi (20″ mouth size) that can contain about 20 L of water (Fig. 8). In the following early morning between 5 a.m. and 8 a.m. farmers release fertilized eggs in the preselected hatching bundh. Farmers release one handi of eggs in a 23 × 15 × 4 ft size bundh, 1.5 handi of eggs in a 38 × 13 × 3 ft sized bundh and half a handi of eggs in a 12 × 7 × 4 ft sized hatching bundh. Within 20 min after release of eggs to the hatching bundh, they supply water mixed with special soil and increase the height of the water body up to 4 inches. Then at a regular interval they examine the eggs and wait until the embryos exhibit quivering or twitching movements. The next morning from 9 a.m. onwards the breeders enter into the bundh and walk normally through the water at an interval of 1 h to prevent settlement of eggs and to maintain temperature to avoid excess heating of the upper layer of the water body (Fig. 10). But in the months of May and June from 10 a.m. to 3 p.m. farmers make a shade from palm leaves (Fig. 11) to avoid excess heat. When the embryos exhibit rapid twitching movements, the fish breeders supply more water to the bundh to complete hatching. It takes 6–7 h for completion of hatching practice after releasing of eggs to the hatching bundh. It is noted that to get maximum results in

Figure 10 Movement of water in hatching bundh by farmer's legs to prevent settlement of egg.

Figure 11 Farmers providing shades to the hatching bundh with palm leaves for temperature regulation.

hatching in this place the congenital temperature for hatching of IMC should be 28°C in water and 42°C in air. The fish breeders start preparations of the hatching bundh for the second operation immediately after the completion of the first operation (Fig. 12).

7. Rearing of hatchlings

The hatchlings or spawn remain in the same hatching bundh for 2 days. Only the water temperature is controlled by constructing a shade upon the bundh by palm leaves (Fig. 11). The water supply is necessarily twice in a day to maintain the height of the

Figure 12 Hatching bundh filled with water after cleaning for reuse.

Figure 13 Examination of fertilized eggs to assess the approach of hatching time.

water bodies. During rearing the fish breeders make a regular examination of the developing hatchlings (Fig. 13).

8. Collection of hatchlings

From the third day onwards since hatching, the hatchlings from different hatching bundhs are collected and transferred into hatching hapa. The size of this earthen hapa is 18 × 10 × 4 ft, but it also varies depending upon the availability of the hapa. In the premonsoon period the carrying capacity of the hapa is 30 bati

(135 mL/bati), whereas in the monsoon period it is 40 bati. (One bati contains about 40,000–50,000 spawn at 3 day old.)

9. Production of hatchlings

     In the premonsoon period hatchling production is 9–10 bati (135 mL) (Fig. 14) per handi of eggs, whereas in the monsoon period the production is much higher than the premonsoon period (about 10–12 bati per handi of egg).

10. Marketing of spawn or hatchlings

Figure 14 Counting of spawn by using a conventional bati (spawn-holding capacity known) for selling.

Figure 15 Oxygen packing of spawn/hatchlings for transportation.

On the third (in hatching bundhs) afternoon the hatchlings are counted by special bati (Fig. 14) after collecting it from hapa and packed with oxygen into plastic bags for marketing. The packaging of 2-day-old hatchlings or spawn (size about 5-6 mm) of Panchmura area is made either by oxygen packing in plastic bags (Fig. 15) (each bag can hold one bati of spawn) or by aluminum handi (each handi contains one bati of spawn).

Hybridization in Bundhs

In "bundhs" breeding occurrence of natural hybrids may be produced due to congestion in the spawning ground. Because of the limited space, there is every likelihood of ova of one species being accidentally fertilized by the sperm of another species. Particularly in the case of IMC with very compatible genomic structure (Zhang and Reddy, 1991), it is all the more easy and frequent to encounter natural hybrids when bred together in a relatively congregated condition.

References

Alikunhi, K.H., Vijayalakshman, M.A., Ibrahim, K.H., 1960. Preliminary observation on the spawning on Indian carps, induced by injection of pituitary hormone. Ind. J. Fish. 7 (1), 1–19.

Alikunhi, K.H., Sukumaran, K.K., Parameswaran, S., Banerjee, S.C., 1964. Preliminary observations on commercial breeding of Indian carps under controlled temperature in the laboratory. Bull. Cent. Ind. Fish. Res. Inst. Barrackpore 3, 20.

Atz, J.W., Pickford, G.E., 1959. The use of pituitary hormones in fish culture. Endeavour 18 (71), 125–129.

Banerjea, S.M., 1967. Water quality and soil condition of fish ponds in some states of India in relation to fish production. Indian J. Fish 14 (1 and 2), 115–144.

Chattopadhyay, N.R., Ghorai, P.P., De, S.K., 2013. Bundh breeding-rejuvenation of a novel technology for quality seed production by the fish seed producers of Bankura District in West Bengal, India. Inter J Current Sci 9, 123–132.

FAO, 1997. Review of the state of world aquaculture. FAO Fish. Circ. No. 886, Rev. 1, 163 pp.

Fermin, C.A., 1991. LHRHa and domperiodone-induced oocyte maturation and ovulation in bighead carp, *Aristchthys nobilis* (Richardson). Aquaculture 93, 87–94.

Hora, S.L., 1945. Symposium on the factors influencing the spawning of Indian carp. Proceedings of the National Institute of Science of India (B) 11, 303–312.

Jhingran, V.J., 1991. Fish and Fisheries of India. Hindustan Publications (India), Delhi, 954 pp.

Khanna, D.V., 1958. Observations on the spawning of the major carps at a fish farm in the Punjab. Indian J. Fish. 5 (2), 282–290.

Khanna, S.S., 1992. An Introduction to Fishes. Central Book Depot, Allahabad.506–512

Lin, H.-R., Van Der Kraak, G., Zhou, X.-J., Liang, J.-Y., Peter, R.E., Rivier, J.E., et al., 1988. Effects of [D-Arg6, Trp', Leu', Pro'NEt]-luteinizing hormone-releasing hormone (sGnRH-A) and [D-Ala6, Pro'NEt]-luteinizing hormone-releasing hormone

(LHRHA), in combination with pimozide or domperidone, on gonadotropin release and ovulation in the Chines loach and common carp. Gen. Comp. Endocrinol. 69, 31–40.

Padhi, B.K., Mondal, K., 1999. Fisheries Genetics. Oxford- IBH, New Delhi.

Sundararaj, B.I., Goswami, S.V., Donaldson, E.M., 1972a. Effect of salmon gonadotropin on in vitro maturation of oocytes of a catfish, Heteropneustes fossilis. J. Fish. Res. Board Can. 29, 435–437.

Sundararaj, B.I., T.C. Anand, and V.R.P. Sinha (1972b) Effects of carp pituitary fractions on vittellogenesis, ovarian maintenance and ovulation in hypophysectomized catfish *Heteropneustes fossilis* (Bloch).

Zhang, S.M., Reddy, P.V.G.K., 1991. On the comparative karyomorphology of three Indian major carps, *Catla catla* (Ham.), *Labeo rohita* (Ham.) and *Cirrhinus mrigala* (Ham.). Aquaculture 97, 7–12.

INDUCED BREEDING— A SCIENTIFIC APPROACH TOWARDS MODERN FISH BREEDING PROCEDURE

The implementation of scientific breeding procedure by fishermen goes way back since they first introduced bundh breeding. However, problems were encountered by the fishermen as the bundh breeding of fishes were erratic and in 1960s, the spawn production industry of the districts faced a serious problem. It was reported by the local fish farmers of Bankura that a large number of bundhs turned sterile where the fishes did not response to breed even when possible congenial conditions were created. Hence further scientific approach in bundh breeding was taken up in the year 1968 when a group of scientists from the Fisheries Research Station, Kulia, Kalyani, introduced the system of induced breeding by application of hormone injections to 10–15% of fishes of the bundhs. This introduction of injection system to fishes gave a new dimension to the spawn production industry of the districts and also opened a new avenues for breeding fishes by sympathetic induction, that is, when injecting a part of the brood stock the rest of the stock is induced being influenced by the induced stock. In spite of immense economic potentialities, few attempts have been made to improve upon existing methods of artificial induction of spawning in fishes. Our study indicate, the fish breeders of India practicing the same technique developed nearly 60 years ago by the pioneers Prof. B.A. Houssay, University of Buenos Aires, and Dr. Von Ihering, Department of Fisheries, Brazil. Even in the most developed countries, there exists a compartmentalization between theory and practice. In reality, it is observed that the investigator is either a fish physiologist or endocrinologist, not conversant with the practical aspects of fish breeding and behavior. Similarly the fish breeders at field are not aware of the scientific information of breeding, spawning, and hatching. Now, to make the seed production industry more profitable, an integrated approach is needed.

ECO-HATCHERY FOR FISH BREEDING OF CARPS IN CAPTIVITY

Cement circular hatchery, also known as eco-hatchery and Chinese hatchery system, comprises of an overhead tank, spawning pool, incubation and hatching tank, and spawn collection tank.

1.1 Overhead Tank

The overhead (double chambered) tank is made of R.C.C. and has a capacity of about 30,000–50,000 L. It is used to supply sufficient water for spawning to the incubation and spawn collection tanks. (Fig. 1.1). Water is poured into the tank by two separate G.I. pipes of 7.5 cm. diameter, for two separate tanks (one for natural surface water and another for ground water) inside one tank. The pond water before storing is properly filtered, but in the case of ground water it is oxygenated by a mechanical air-diffusion method. After mixing the water, in the ratio of 1:1, it becomes more suitable for spawning.

Induced Fish Breeding. DOI: http://dx.doi.org/10.1016/B978-0-12-801774-6.00001-8

Figure 1.1 Overhead tank of 30,000–50,000 L capacity.

1.2 Spawning Pool

The spawning pool was evolved in China in 1970 and thereafter introduced to different countries of the world. It is a circular cement tank of 8–9 m diameter and 1–1.5 m depth with 50 cubic meters of water holding capacity. A spawning pool is composed of two parts: (1) a circular breeding pool (6–8 m dia. and 1.2–1.5 m height) and (2) a rectangular spawn collection chamber (3 m × 2.5 m × 1.6 m) located at the base of a circular pool adjacent to it and connected to the pool by a 10 cm diameter G.I. pipe. The bottom of the pool has a slope gradient of 1:30 toward the central aperture, called the *orifice*, through which the fertilized eggs are transferred automatically to the spawn collection tank or directly to the incubation pool. During transfer the orifice is guarded by a 25 mm meshed wire screen to provide easy passage of fertilized eggs but at the same time prevent the escape of breeders. The wall of the tank is provided with diagonally fitted duck-beak inlet for circular water flow to create riverine fluviatile conditions, as a result the carps are stimulated, engage in sex play, and ultimately spawn. About 70 kg of brooders (2 males:1 females) can be used at a time, which will yield about 8–10 million fertilized eggs in one operation. The outlet pipe (Fig. 1.2), from initiation of breeding till spawning, remains fixed to the central orifice in an erect posture so that excess water will be removed from the pool. As soon as spawning is completed the brooders are removed and at the same time the outlet pipe is dislodged (Fig. 1.2). The central orifice is covered by a 25 mm meshed wire screen which allows the fertilized eggs to incubation pool but not the brooders. In order to increase the dissolved oxygen content in the tank a perforated galvanized iron pipe

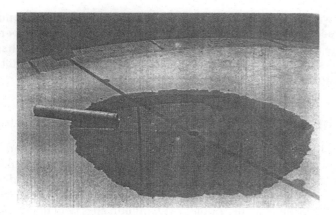

Figure 1.2 Central orifice for automatic transfer of fertilized eggs to hatching pool.

Figure 1.3 Breeding pool in operation with showering in the pool.

is fitted above the breeding tank in such a manner that a fine shower falls inside the breeding pool (Fig. 1.3). The water holding capacity of the circular pool is 35–70 m^3 at its maximum level, but only 20–30 m^3 water is normally maintained for breeding 25–40 sets of brooders at a time. The water inlet to the pool is a 10 cm diameter G.I. pipe emerging out from the overhead tank, which branches into three 5 cm diameter pipes that enter into pool through three separate entry points, and is guarded by a screw valve to maintain clockwise movement of water inside the pool. These pipes are placed 15 cm above the bottom to maintain a steady flow of water at the entire column of water and a flow rate of 30–50 L/min is maintained by adjusting the screw valve.

Nowadays, fish breeders have rejected the concept of conducting breeding operations in a breeding pool to avoid the involvement of huge amounts of water, although some are of the opinion that the central sloping prevents a thorough mixing of all eggs with milt and this reduces the fertilization rate.

1.3 Incubation or Hatching Pools

The incubation or hatching pools are circular in shape and constructed of brick and cement. Generally, 2–3 incubation tanks are connected to one spawning pool (Fig. 1.6). Each has an inside diameter of 3–4 m and a depth of about 1 m with a water holding capacity of 9–12 m³. It usually holds about 0.7–1.0 million eggs per cubic meter of pool. The incubation pool is directly connected to the overhead tank by a 7.5 or 10.0 cm diameter G.I. pipe, which after being branched into two 5.0 cm diameter pipes joins two 5.0 cm diameter circular pipes fitted below the cemented bottom of the outer chamber. Finally, the base line remains attached to both the inner and outer circular pipe from which 21–24 duck-beak inlets of 2.0 cm diameter, are projected above the bottom at a 90 degree angle at regular intervals and in a clockwise fashion. All this arrangement of pipe is to create a centrifugal force of water toward the inner chamber through the monofilament markin cloth and out through the outlet pipe.

Water circulation is very important for the proper hatching of fertilized eggs and the said arrangement of duck-beak inlets creates a three directional flow (Fig. 1.4) at the bottom which ultimately creates a centrifugal force. This three directional water flow does not allow any settlement of eggs at the bottom. The water inlets at the bottom

Figure 1.4 Three directional water flow in hatching pool through duck-beak inlet as observed in the spawning or hatching pool of three states.

and the sides of the incubating pools are technically arranged so as to create a circular movement of water (Fig. 1.5). This keeps the developing eggs in a constant circular movement. The inner chambers of the incubation tanks are separated by a fine monofilament markin cloth (Fig. 1.6). This nylon screen is stretched and fitted on an iron frame. A rubber belt is fastened very tightly to seal the compartment. In the center is the overflow pipe at a particular level below the level of inner chamber through which excess water is led out; as a result a constant water level is maintained in the inner chamber. The water speed is controlled by means of a gate valve. The desired flow rate of 45–60 L/min is an important criteria for the steady development of embryos and also hatching.

Figure 1.5 Enlarged duck-beak inlet, projecting slightly from the wall. The pipe is compressed on both sides to maintain a three directional flow.

Figure 1.6 One breeding pool and three adjacent incubation pools.

1.4 Breeding Technique

First of all, the breeding tank is filled with water. All the diagonal inlet pipes are kept open to maintain a constant circular motion inside the breeding pool. After injection, the spawners and milters are released in the breeding pool. Due to the constant circular movement of water, fluviatile conditions are simulated. The brood fish, on being stimulated, perform breeding following sex play. The fertilized eggs are led into the incubation chambers through a central outlet pipe which is controlled with a gate valve. The flow of water in the incubation chambers is controlled to keep the eggs constantly in a circular motion due to diagonally arranged pipes. Technically, such an unique ordering of pipes and maintenance of constant water level by arranging easy outlet from the inner chamber facilitates the development of fertilized eggs. The eggs, through the process of water hardening, gradually transform into a pearl-like structure from a pinhead. At this stage the egg is so transparent that the developing embryo is visible within it. The excess water is led out through a central outlet pipe fitted in the center of the inner chamber. The spawn is retained for three days in the pool in flowing conditions, until the yolk sac is absorbed.

1.5 Spawn Collection Tank

The spawn is led into the spawn collection tank, which is rectangular in shape, in which a fine nylon hapa is fixed. The spawn gets collected in the nylon hapa from where it is counted using bati (a small bowl whose holding capacity of spawn is known) and transferred to the nurseries (Fig. 1.7).

Figure 1.7 Spawns are counted with the help of a mug before sale.

1.6 Breeding Protocols

Induced breeding of fishes starts with the collection and selection of the correctly matured brood stock prior to the breeding season and their subsequent management in brooders ponds. The fish breeders are often confused about the proper management and being convinced, by the agents and/or dealers, that more feeding will enhance early maturity along with greater numbers of eggs and milt, advocate feeds much greater than that of the required optimum amount. This results in the formation of a dense scum on the surface with resultant supersaturation of oxygen, leading to mass mortality of brooders particularly during afternoon, indicated by surfacing behavior. Scientific counseling for proper management of brooders is still absent from the study area and the farmers, both in the freshwater and brackish water sectors, are guided (misguided?) by the agents and dealers of unauthorized feed, fertilizer, medicine, and drug companies.

1.7 Brood Stock Collection and Development

The collection of brood stock in adequate number from different geographical territories, cataloguing of their geographical origin, their genetic characterization, and the maintenance of pedigree records are important prerequisites for a scientific breeding program. These aspects are of much genetic relevance where brood stock collection and management is concerned. In addition, qualitative feeding with the right and required ingredients along with periodic health management are important aspects should be undertaken with the utmost care. Depending on the available criteria, fishes exhibit considerable variations in the number and quality of eggs that develop, i.e., fecundity varies. As fecundity refers to number of eggs/kg of body weight, it is easy to assess the feed requirements of brood stock for the production of a specific number of eggs. Experimental results indicate egg production is reduced by 75% when ration size is reduced by half, further decreases in ration size during the second half of the reproductive cycle reduces egg size.

Brood stocks may be comfortable on maintenance rations but, as revealed by various experimental results, deficiencies of certain dietary ingredients, such as fatty acids (PUFA), vitamins, and trace elements, exerts a negative impact on maturation, breeding, spawning, larval vigor, and survival. As we know that nutritional requirement varies according to species, proper experiments should be designed to formulate the right food for individual brood stocks so that the supply of quality seed to the farming sector is ensured.

Hatcheries are categorized into monospecies, multispecies, indigenous carp hatchery, exotic catfish hatchery, prawn hatchery, and so on depending on the species considered for breeding. As the species separation is not so stringent in fish, unlike other vertebrates, and as the genetic flexibility is extended up to generic and higher taxa, the breeding of different stocks should be conducted separately to avoid uncontrolled hybridization. Nowadays, our better understanding and the development of specific technologies can help us to understand how proper brood stock management is important for quality seed production and why it has been given priority in hatchery management.

Generally brood stock management program is divided into two broad categories:

(1) The prespawning and (2) the postspawning management program. The prespawning program includes procedures such as selection of suitable brood stock and procurement, acclimatization, maintenance, maturation, spawning, and hatching. While postspawning management includes facility maintenance, water quality management, brood stock selection, health management, risk assessment, health management, documentation, and record keeping.

Although selective breeding and hybridization programs of pedigreed fish are one of the primary criteria for quality seed production in captivity, they are not carried out in most hatcheries. Generally brood fish are collected either from the wild or cultured stocks in neighboring farms or hatcheries, or by developing new brood stocks through random selection from previous fish stock. One of the most common problems in selecting broods from rearing ponds is that only undersized and undergrowth fingerlings are left for selection, since the table-sized fish have been sold. Therefore, positive mass selection, involving individuals with required cultural traits and free from external deformities will not be available. Nevertheless, the practice of sex-wise segregation, which is an important consideration, is ignored throughout the program by all categories of hatcheries due to lack of sufficient infrastructure, ignorancy, and due to a lack of knowledge of the consequences. The good practice of transferring spawners after spawning to a resting pond and not keeping them with ripe males and females is ignored too in most of the hatcheries. The improvement of domesticated brood stock may be initiated through specific genetic improvement programs (GIFT, GET, EXCEL, GST). The development of GIFT tilapia demonstrates how genetic improvement can be done by the application of genetic improvement tools. In Bangladesh, Philippines, Thailand, and Vietnam, national breeding programs and tilapia genetic research are based mainly on GIFT or GIFT-derived strains. GIFT tilapia is now extensively used as a ideal basis for the development of tilapia farming in several South Asian countries. GIFT tilapia improvement procedures are now in use for the genetic improvement of species like silver barb in Bangladesh and Vietnam, roué in Bangladesh and India, mrigal in Vietnam, and blunt snout bream (*Megalobrama amblycephala*) in China.

Several brood stock management issues are being experienced by the fish seed industry. Ingthamjitr (1997) reported that hybrid catfish farmers in central Thailand had suffered a decline in production of good quality seed instead of sound management practices. As the introduction of fresh germplasm at certain intervals either from nature or from distant geographical territories is not practiced in any hatcheries, this is believed to be the main reason for depression in growth in India (Eknath and Doyle, 1985). Very often environmental issues and management of quality seed production is ignored by the fish breeders and the husbandry managers as well. However, the current initiatives of stock upgradation will not be sustainable unless the scientific reasons are better understood and specific improved management strategies are developed (Little et al., 2002).

To avoid potential problems relating to heredity, impaired growth and poor survival due to inbreeding. and other related consequences, the pedigree record of different family lines of domesticated stocks (foreign or native) must be in the custody of management. Again, the performance and development data of the candidate families or lines under a uniform range of environmental conditions is needed. Selection protocol is also an important criteria, i.e., whether the stocks were selected from populations having better performance and survival following disease outbreak and the exact timing of the selection procedures. Large as well as government hatcheries are not practicing record keeping, which may be due to carelessness, ignorancy, and not being appraised of the negative consequences. Hatchery operators, at all levels, should be made aware of the importance of record keeping and the category and source of brood stock. A system of brood stock certification will be an added advantage in any improvement program.

As we know that the quality improvement of brooders is directly related to the nutritional requirement, proper high nutrient feed with balanced ingredients should be provided. Tilapia females should be provided high nutrient feed after their incubation period as they starve for the entire period (10–13 days). Initiatives should be directed at the development of low cost feed with the desired nutrient quality. Therefore, alternative feed sources or an appropriate formulation for farmer-made aqua-feeds should be made available for small-scale farmers.

One unavoidable criteria strikingly absent from brood stock management practice is quarantine aspects. Quarantine facilities are essentially a closed holding area where brood fish are kept in individual tanks until the results of screening for known diseases or disorders are identified. Such facilities can be afforded by most large-scale and government hatcheries but are not well maintained. Quarantine holding units should ideally be placed at a distance from the seed production unit. If they are not, measures should be taken to ensure that there will be no contamination. Currently, harmonized technical standards for the production of regional fish seed in fresh water sector are lacking. There is a current need to develop such technical standards with subsequent standardization and validation at the national and international level, specifically for those species cultured in diverse geographical territories.

1.7.1 Gene Banking

The development of gene banks to improve the genetic quality of brood stock and thus the quality of seed is considered as an important tool for quality seed production in the aquaculture sector. While gene banking now is almost a standardized procedure for upgradation of genetic stock in the agriculture and livestock sectors, fish gene banks are rare and still not supported adequately, especially in developing countries (Brian et al., 1998). One of the most valuable fish gene banks in Asia is the Nile tilapia brood stock assembled for the development of GIFT, together with the GIFT synthetic base population and subsequent generations of selectively bred GIFT in the Philippines at the National Freshwater Fisheries Training Centre (NFFTC). The descendants of these fish remain available from the said gene bank for national, regional, and international breeding purposes.

Considerable progress has been made in Bangladesh with respect to gene banking. From 2002 to 2003, the Department of Fisheries, Bangladesh initiated the establishment of 12 fish brood stock banks in the Government Fish Seed Multiplication Farms with a target production of 110 tonnes of genetically improved brood of Chinese carps, 1800 kg spawn, and 0.5 million fingerlings. Fries for brood stock banks are collected from different rivers. To date, 85 tonnes of brood have been produced and distributed to public and private hatcheries under the newly formulated fisheries policy. Another 20 brood banks have been established in 20 Fish Seed Multiplication Farms and one Fish Breeding and Training Centre under the Fourth Fisheries Project. Moreover, an NGO (BRAC) has established one carp brood bank and is planning to set up another. These brood banks provide the necessary training for other government and private hatchery operators on brood stock management. Although a cryogenic gene bank has not been established as yet, research is in progress in the cryopreservation of sperms of Indian major carps (*Catla catla, Labeo rohita, Cirrhinus cirrhosus*) and Chinese carps (*Cyprinus carpio, Ctenopharyngodon idella, Hypophthalmichthys molitrix, Aristichthys nobilis, Barbonymus gonionotus*).

Experimental studies of wild stocks of *C. catla* have shown that they represent a diversified genetic resource and indicates that in situ management practices, such as preventing the wanton capture of fish and creating sanctuaries for protecting small stocks such as those in the Halda river, can help maintain and conserve the present diverse gene pool. Hatchery owners are accustomed to operating negative selection and polygynous breeding systems in which some males mate with many females year after year (Islam et al., 2007), resulting in genetic deterioration that subsequently causes a negative impact on aquaculture production. The current research advises hatchery owners to collect or replace broods from genetically developed and diverse fish from the stocks of the Halda and Jamuna rivers. This will increase the effective breeding populations and thus help to improve the aquaculture production. However, the strict implementations of correct management practices are essential for the maintaining genetic diversity of the natural stocks. This practice of breeding between two genetically discrete or distant populations would certainly exert a positive impact on quality production in aquaculture.

Brood stock management is an important part of general aquaculture practice and interrelated to all other segments of aquaculture production cycle. It is however often considered to be difficult by some hatchery operators due to their lack of knowhow or simply overlooked by others. The issue is further complicated by lack of overall planning, little collaboration among seed producers, insufficient financial input for R&D, and lack of institutional support. Efforts to maintain and improve the brood stock quality of any major cultured species requires long-term strategic planning at national and regional levels and practical approaches involving public sectors, breeding centers, and private hatcheries at various operational scales. Capacity building for all stakeholders through training is therefore fundamentally important to raise awareness, update knowledge, and enhance skills.

References

Brian, H., Ross, C., Greer, D., Carolsfled, J. (Eds.). 1998. Action before extinction. In: An International Conference on Conservation of Fish Biodiversity. World Fisheries Trust, Victoria, BC.

Eknath, A.E., Doyle, R.W., 1985. Indirect selection for growth and life-history traits in Indian carp aquaculture. 1. Effects of broodstock management. Aquaculture 49, 73–84.

Ingthamjitr, S. 1997. Hybrid catfish Clarias catfish seed production and marketing in Central Thailand and experimental testing of seed quality (PhD Dissertation). Bangkok, Asian Institute of Technology. 135 p.

Islam, M.N., Islam, M.S., Alam, M.S., 2007. Genetic structure of different populations of walking catfish (*Clarias batrachus* L.) in Bangladesh. Biochem. Genet. 45, 647–662.

Little, D.C., Satapornvanit, A., Edwards, P., 2002. Freshwater fish seed quality in Asia. In: Edwards, P., Little, D.C., Demaine, H. (Eds.), Rural Aquaculture. CABI Publishing, Oxon,United Kingdom, pp. 185–193.

2

REPRODUCTIVE CYCLE, MATURATION, AND SPAWNING

For a successful induced breeding program, a clear understanding of the reproductive cycle, maturation, and spawning is of the utmost important. In most fish spawning is characteristically an annual event and the sperm and eggs are released at a time of the year when the conditions of spawning are in such a favorable condition that, although fertilization is external, it supports maximum survivality by providing ambient environmental conditions and natural food. Most

Induced Fish Breeding. DOI: http://dx.doi.org/10.1016/B978-0-12-801774-6.00002-X

fish use seasonal patterns of changing day length, temperature, food availability, and rainfall, in particular, to adjust their maturation and for successive spawning and hatching. Some of the earlier studies indicated that gonads start enlarging at a time when day length and temperature are in an increasing trend. Of the various environmental factors day length and temperature have been singled out as the most important factor in regulating reproduction in freshwater fish. The temperature and food supply may vary from year to year, as is happening now due to global warming, but at a particular latitude a given day always has a certain length, which is longer or shorter, by a given interval, than the preceding day. This indicates that the seasonal increase or decrease of day length has a constancy which other environmental factors lack. For example, in the northern hemisphere, the shortest day is on December 22 (winter solstice), thereafter the day length increases until it reaches the maximum on June 21 (summer solstice). Thereafter, the day length decreases again. On 2 days in a year, namely, March 20 and September 22, the day and night are of equal duration. The reverse is true in the southern hemisphere. The gonads of most freshwater fish enlarge precisely at a time when the day length increases beyond 12 h/day after the vernal equinox (March 20). This mechanism ensures gonadal enlargement prior to onset of monsoon rain, which induces spawning. Fish, like most other animals, use this as a reliable cue for the timing of their gonadal recrudescence in as much as seasonal day length variations are very precise and predictable, although it is not certain how fish feel variations in day length.

Experiments showed that exposure of fish to artificial illumination for 14 h/day during January advances the formation of large eggs by 3 months, and such precociously gravid females have been successfully spawned by injecting pituitary extract. Again, when the spent fish are exposed to long days, the fish develop a second crop of eggs within 1 month. Exploring this trick, fish are now made to spawn successfully four times between March and July. This exhibits that it is possible to harvest four crops of eggs in 5 months, which is not possible in natural spawning. In fish that are stocked in impounded waters, often the peak of reproductive activity of both the sexes does not coincide. Males usually attain maturity much earlier than females, and the reverse may happen in a few cases. In either case fertilization is impaired, resulting in reduced production of spawn. This uneven attainment of maturity can be made to synchronize by exposure to the required artificial illumination for a specific time frame. Other experimental observations indicate promising results which may revolutionize seed production in aquaculture. When one of the ovaries is removed, the remaining ovary not only enlarges but doubles its component of eggs within 3 weeks. This offers an exciting avenue to earn more revenues for those who harvest fish for eggs and making caviar (salted sturgeon roe or ovary).

In addition, to enhance maturity, reproduction often is accompanied by reorganization of somatic and reproductive tissues and development of marked secondary sexual characteristics, which include increased water accumulation with resultant loosening of muscle, markedly of the region in and around gonads. Along with this, flesh becomes less palatable due to an increased loss of protein and fat and is evidenced by the increased availability of fat and protein in the water of brooders ponds. Other changes include darkening of skin, roughness of the dorsal surface of the pectoral fin of male carp, and protrudable upper or lower jaw, known as kype, in male salmonoids. One of the remarkable phenomena along with maturity is the increased susceptibility of the brooders to secondary infection by fungi and bacteria through the lesions developed during courtship, in the case of catfish, salmon, pacu, and many other fish which exhibit aggressive behavior. Along with this, the trauma of breeding and spawning produce a killing effect in males, but the female is found to be ready for the next spawning. This is observed in Pacific salmon, trout, eel, etc.

2.1 Collection and Preservation of Pituitary Gland

One of the prerequisites for an induced breeding program is the pituitary gland. The gland is generally collected by the gland collector from the ice-preserved beheaded fish, from the market. For collection, a hole is scooped out from the rear end of the head just below the brain, along the sphenoid bone. After that the brain is carefully scooped out from the hole, after the removal of the brain, the gland is clearly visible, lying in the cavity known as the sella turcica (Fig. 2.1).

The muscles attached to the glands are removed carefully by moving a needle around the gland in the cavity. When the gland is free then it is taken out with the help of a spoon and kept in absolute alcohol for preservation (Fig. 2.2). The gland collectors sell the glands to the retail suppliers who preserve it in a refrigerator maintained for this purpose. During the breeding season the fish breeders purchase bulk amounts of these glands, as per the needs of the day-to-day program, from the retail suppliers at the rate of $3–5 per piece. The weight of the gland generally varies from 3 to 6mg and it is preserved in absolute alcohol with two changes in 24h, and finally kept in amber-colored phials with absolute alcohol. The fish breeders reported that the gland can be preserved in refrigerators for up to 2 years by changing the alcohol from time to time. During the breeding season, the fish breeders temporarily

Figure 2.1 Pituitary gland taken from the sella turcica below diencephalon of the hypothalamus of the brain.

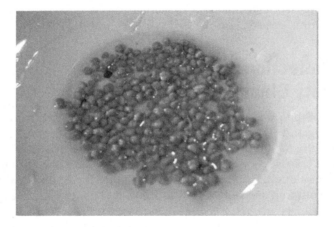

Figure 2.2 Pituitary glands on blotting paper for absorption of alcohol.

preserve the gland by placing it in a dark bottle with absolute alcohol and by placing the bottle in wet sand (Fig. 2.3). It may be noted in this regard, as the pituitary gland is collected from ice-preserved fishes, that the potency of the gland is reduced and this may lead to a reduced population, which ultimately may result in inbreeding. The fish breeders do not have any idea about these negative aspects of pituitary gland collection.

The extract of the gland is usually prepared just before injection. At first the gland should be completely dried by placing it on blotting

Figure 2.3 Gland in a dark bottle kept in wet sand for long-term preservation.

paper (Fig. 2.2), as the presence of a trace amount of alcohol in the gland may produce lesions to the fish. The gland is then weighed individually in a simple pan balance and homogenized in distilled water, using tissue homogenizer (Fig. 2.4A). After that the homogenized gland tissue is centrifuged for around 5–7 min to allow precipitation of ground gland tissue at the bottom of the tissue homogenizer (Fig. 2.4B). As a result, a clear Gonadotrophic hormone (GTH) in dissolved water, known as supernatant, is available over the precipitated gland tissue. The supernatant fluid in which GTH hormone exists in dissolved form is decanted from the centrifuge tube and used for injection. During bulk operation distilled water is added to the supernatant in such a way that the desired volume is prepared for the number of fish considered for the breeding program, generally the extract dose is limited to 0.5–2 mL/kg body weight, but the dose should not exceed 1–1.5 mL in any case.

2.2 Identification of Male and Female Breeders

Fish are sexually dimorphic, i.e., they appear in two different sexes, but there are very few secondary sexual characters in most fish to differentiate their sexes morphologically (Table 2.1). In carp, the secondary sexual characteristics are not very prominent and conspicuous, by which the sexes of the race can be distinctly identified. Some morphological characters, like roughness of the pectoral fins, scales, operculum, etc., developed during the breeding season in

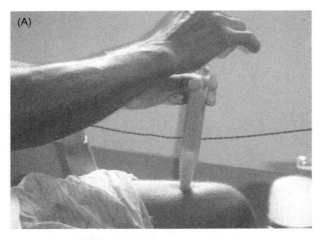

Figure 2.4 (A) In-field homogenization just before induction. (B) Inner view of electrically operated field centrifuge showing four holders for holding centrifuge tubes and for centrifugation.

both sexes of carp can, however, be used as a key to identification of sex. Morphological differences during the breeding season are more or less the same in all species of various genera of carp.

The mature male can be identified by the roughness of the dorsal surface of the pectoral fin, which contrasts with the very smooth fin surface in the female (Fig. 2.5). The roughness is felt by touching the surface of the fin close to the body. This roughness in the male fish may appear as early as March, but it disappears when the breeding season is over. The ripe male and female fish can also be distinguished by the shape of the body, condition of the vent, and the secretion of milt (milky white secretion) in males (Fig. 2.6A). A fully

Table 2.1 The Common Characteristics for Identifying the Sex of Mature Indian Carp

Sl. No.	Characters	Male	Female
1	Scale	Rough with sandy texture, especially on the flanks and anterior dorsal side and nape	Scales smooth and silky
2	Operculum	Rough with sandy tubers	Operculum smooth
3	Pectoral fins	Rough with sandy touch, particularly on the dorsal side. Pectoral fins slightly stouter and longer in a majority of Indian carp	Pectoral fins very smooth and slippery, smaller in length in comparison (Fig. 2.5)
4	Abdomen	Round and firm and not very soft to the touch	Bulging out on both sides. Soft and palpable. A distinct cleavage is formed midventrally along the abdomen when the fish is placed on its back
5	Vent	Elongated slit introverted (concave); white in color	Round slit extroverted (convex), fleshy and pinkish in color, papilla prominent
6	On pressure on abdomen	On slight pressure above the vent on the abdomen, milky white fluid runs out through the vent	On pressure, slight yellowish discharge or a few ova may come out through the vent, in very mature condition; ova would slip out on slight pressure

Figure 2.5 Mature brooders of *Pangasius*, female with swollen abdomen and smooth pectoral fin, while the male has a comparatively large and rough pectoral fin along with a protruding vent.

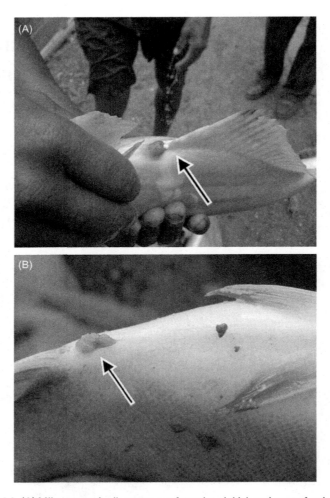

Figure 2.6 (A) Milt automatically oozes out from the pinkish-red vent of a ripe male, with a large and rough pectoral fin. (B) Eggs automatically ooze out from the reddish vent (not as protruding as in the male) indicating the female is in the peak of maturity.

ripe gravid female has a bulging abdomen with a swollen pinkish-red vent. In the ripe male, the abdomen is almost flat and the vent protrudes like a tube. The gravid ripe females are sometimes found with a flat vent through which the eggs ooze out through it easily (Fig. 2.6B). This condition of the female breeders is usually found in fish collected from rivers or reservoirs. The ripe males, which ooze milt through the vent on slight pressure on the abdomen, can be segregated easily from the female breeders.

Table 2.2 The Length of the Pectoral Fin of Some Indian Carp (After Choudhuri)

Sl. No.	Species	Pectoral Fin Reaching the Lateral Line Scale	
		Male	**Female**
1	Catla (*Catla catla*)	8th or 9th	6th or 7th
2	Rohu (*Labeo rohita*)	8th or 9th	6th or 7th
3	Mrigala (*Cirrhina mrigala*)	10th or 11th	8th or 9th
4	Calbasu (*Labeo calbasu*)	10th or 11th	8th or 9th
5	Gonius (*Labeo gonius*)	18th or 19th	16th or 17th
6	Bata (*Labeo bata*)	8th	7th
7	Reba (*Cirrhina reba*)	8th or 9th	6th or 7th

Identification of male and female breeders of the major and minor Indian carp can also be done during the breeding season from the nature and length of their pectoral fins (Table 2.2). The fins in mature males of the majority of species are slightly stouter and extended backward and toward the dorsal side of the lateral line scales at a place more posterior than that in mature females.

2.3 Selection of Breeders

Success of inducing breeding of fish depends largely on proper selection of suitable breeders. The selection of good male breeders is easy, as a fully ripe male fish oozes milt freely on being gently pressed at the belly (Fig. 2.6A), but the selection of suitable female breeders is difficult. Female breeders having a soft, more rounded, bulging abdomen with swollen, reddish vent are selected for the purpose (Fig. 2.6B). The right indication is the oozing out of individual eggs one by one through the vent on application of slight pressure on the abdomen behind the vent portion. It is preferable to choose healthy breeders in the age group 2–4 years with their weights ranging from 1 to 5 kg. The age of the breeders can be known correctly only if the fish are from a known stock.

The approximate age can be determined by examining the scales, size of the fish, etc. Medium-sized fish are easier to handle, and the number of eggs is also large. Larger-sized breeders are not usually

Figure 2.7 Collection of brooders (pacu) in a handbag from a hapa.

selected for the reason that they are difficult to handle and the fish accumulate fat in their body cavity. Besides, more fibrous tissue may develop in the ovary of the female fish, hampering spawning. At the time of selection, the breeders are taken one by one in the handnet and the sex determined quickly by inserting a hand into the net and feeling the pectoral fin by the tip of the fingers. Taking the breeders in the handnet prevents the fish from struggling too much and injured themselves.

The induced-fish breeding experiment begins with the onset of the monsoon when the fish become fully ripe. At the time of experiment the breeders are netted out with a dragnet and the males and females are kept segregated in separate nylon hapa in a nearby tank. At the time of injection the broods are collected by hand and transferred to the site of injection in a soft handbag (Fig. 2.7). The handbag is usually made of 4–6 mm meshed net cloth stitched in the shape of a bag with 70–75 cm length and 50–55 cm breadth. Old pieces of fry net cloth may also be used for preparation of the handbag. This bag is more useful than previously used handnets for safe carriage of the breeders to the site of injection and carrying them back for release in the breeding pool after injection. Holding the breeders with the hand is not advisable and should be avoided as far as practicable. Rough handling may cause injury to the fish.

The breeders are collected from the dragnet one by one in the handbag (Fig. 2.7), so that the fish do not struggle and then the weights of the breeders along with the handbag or using strong wire basket are taken by means of a balance (Fig. 2.8).

The actual weight of each fish is calculated by deducting the known weight of the handnet from the total weight of the fish along

Figure 2.8 Weight of individual breeders is recorded before injection.

with the net. After noting the weight, length, and other particulars of a set, the fish are kept separate in an enclosure of the net while the pituitary extract is prepared

2.4 Mode of Injection/Dosage Administration

Though different methods of injection, like (1) intramuscular, (2) intraperitoneal, and (3) intracranial, are practiced by fish-breeding workers in different countries, intramuscular injections are found to be more commonly given. Brazilian fish culturists largely follow the intramuscular method, whereas the majority of workers in Japan and the US practice the intraperitoneal method. The intracranial method of injection for administering the pituitary extract in fish was the favorable method in the former USSR a few years back, but workers there have now switched over to the simpler technique of intramuscular injection, finding it more convenient. The intramuscular injection is less risky in comparison with the other methods.

Intraperitoneal injections are usually given through the soft regions of the body, generally at the base of the pelvic fin. However, there is some risk of damaging the internal organs, especially the suspended gonads, when administering an intraperitoneal injection to fully mature fish. Moreover, repeated intraperitoneal injections with gland suspensions, which are not usually fitted, may have harmful reactions.

In India, both the intramuscular and intraperitoneal injections have been experimented on in pituitary breeding, but the former method has been found to be more effective and easier to follow. Intramuscular injections are usually administered on the caudal peduncle (Fig. 2.9) or on the shoulder region. The injection should be given in a lateral line between the dorsal fin and on the dorsolateral side. If more than one injection is to be administered, it is better to give the second injection on the opposite side to the first injection. For intramuscular injection, the needle is inserted under a scale, first parallel to the body of the fish and then proceeding into the muscle at an angle of 45°. The most convenient hypodermic syringe used for the purpose is 2 cc with graduations to 0.1 cc. The size of the needle for the syringe depends on the size of the breeders to be injected. The BDH needle no. 22 is conveniently used for 1–3 kg of carp breeders and no. 19 for larger ones. Needle no. 24 can be used for smaller size of carp (Fig. 2.10).

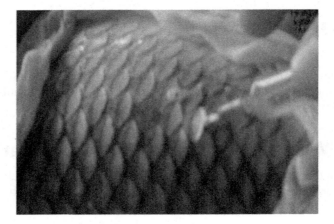

Figure 2.9 Injection by lifting the scale at caudal peduncle.

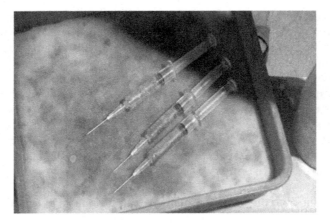

Figure 2.10 BDH needles.

It is advisable not to introduce too much solution for intramuscular injection, as an intraperitoneal injection requires a higher dose of the gland material for successful results. The quantity of solution for intramuscular injection usually varies from 0.1 mL to a maximum of 1.0 mL at a time, depending on the size of the breeders. It is advisable to limit the extract solution to 0.5 mL. The requisite quantity of gland suspension is prepared on the basis of the weight of the breeders to be injected. The required suspension is taken in a hypodermic syringe for injection into the breeders. The breeders in the hand-nets are then brought one by one for injection from the enclosure in a net and placed on a small field table provided with a cushion. Wet pieces of old nets, hapas, cloth, or dunlopillo may be used as a cushion. The field table should be placed by the side of the water where the experiment is carried out. Three persons are required at the time of injection; one holds the head of the fish pressed gently against the cushion, while a second person holds the tail with one hand and with the other hand gives the injection on the caudal peduncle (Fig. 2.9), the third person will readily supply the extract-filled injection tubes one after the other to the injector. The breeders are released immediately after the injection into the breeding pool.

The work, right from placing of the breeders to the injection table, injection and releasing back to the breeding pool, should be done very quickly so that the fish are not stressed, being out of water. Care should also be taken in holding the fish gently but firmly on the table at the time of injection as any sort of pressure on the fish may injure them and ultimately hamper breeding and spawning. There is the danger of the fish giving a sudden jerk, if it is loosely held and this may break the needle and the syringe.

A set of breeders is introduced into each breeding pool. A set usually consists of one female and two males (1:2), depending on the size of the breeders. Sometimes, only one bigger male is introduced with a small female or even three males with a bigger female, if the males are of a small size. Better results are obtained when two male breeders are introduced with one female, because of likely competition among the males in chasing the female. This ensures a high total of spawning, higher fertilization rate, and better chance of synchronization in shedding of milt by the males and of eggs by the female.

2.5 Injection and Dosage for Indian Carp and Exotic Carp

The time of the first injection is shifted to the morning (9–10 a.m.) from the evening, especially in case of stripping for better fertilization and hatching 6 h after the first injection, the second injection is

Table 2.3 Dose of Pituitary Extract for Indian Carp and Exotic Carp

Fish Species	Female		Male
	First Dose	Second Dose	
Indian major carp	1–2 mg/kg body wt	5–8 mg/kg body wt	Single dose 1–2 mg/kg body wt (during second injection to female)
Exotic carp	2 mg/kg body wt	8–9 mg/kg body wt (silver carp)	Single dose 1.5–2.0 mg/kg body wt
		10 mg/kg body wt (grass carp)	Single dose 1.25–2.5 mg/kg body wt (Bata)
Pangasid catfish	1.5 mg/kg body wt	6 mg/kg body wt	1 mg/kg body wt
Pacu (*Piaractus brachypomus*)	2 mg/kg body wt	10–12 mg/kg body wt	2 mg/kg body wt
Magur (African and Thai)	Summer dose 8 mg/kg body wt	Winter dose 10–18 mg/kg body wt	

administered (by 3–4 p.m.), but on cloudy days the second injection is advanced by 1 h. Due to this shift in time, spawning and fertilization occur in the early morning, which provides more a congenial environmental temperature; besides it also removes the hazards of awakening in the night (Table 2.3).

2.6 Three Doses of Injection Instead of the Usual Two Doses

The farmers reported that when the extract is divided into three doses, it ensures better spawning and egg production compared to two doses, and in that case the dose is as follows:

First dose—1 mg/kg body wt
Second dose—2½ mg/kg body wt
Third dose—3½ mg/kg body wt

Second injections are given 3 h after the first and the third dose is given 3 h after the second.

2.7 Calculation of Dose for Purposes of a Batch of Fish

The dose requirement for a 2-kg brood is 2 mg/kg, then the extract, prepared from 4 mg gland, after required dilution (0.5–1.0) is injected into the fish. Similarly, for 50 fish of 2 kg body weight, each requiring a dose of 2 mg/kg, a total of 200 mg gland should be considered for preparation of extract.

2.8 Inducing Agents

Current inducing agents used by the farmers are of two types: (1) natural inducing agents, pituitary extract containing GTH-I and GTH-II, and (2) synthetic inducing agents, Wova-FH, ovaprim, ovatide, etc.

Of these two inducing agents, farmers prefer preserved carp pituitary gland because of its low cost and high success rate, as well as its organic origin. The quality of the gland is assured by the farmers by observing its physical appearance, mainly the color and shape of the gland and also its elasticity. A freshly preserved gland is more elastic and brownish. Contrary to this, it is soft and pale yellow in color if it is not properly preserved. The dose and frequency of injection vary depending upon the species and certain environmental conditions. As per farmers' experience, during windy and rainy days the dose of injection is reduced. Also, a sandy soil bed requires a lower dose for induction. Other than pituitary gland, only a few farmers use ovaprim, ovatide, WOVA-FH, etc., for hypophysation.

2.9 Environmental Influence on Dose

Temperature exerts a profound influence on the doses to apply. Higher temperatures to a certain extent reduce the dose.

2.10 Application of Terramycin With Extract

As stripping induces 30% mortality the fish breeders inject Terramycin (1 mL for 10 kg fish) along with pituitary extract with the understanding that mortality will be reduced to 50%.

2.11 Methods of Injections

Generally injection is applied on two sites as below.

2.11.1 Caudal Peduncle

Most of the farmers administer injection at the caudal peduncle, i.e., the region on the dorsal side between the dorsal fin and caudal peduncle.

2.11.2 Base of Pectoral Fin

This is not practiced by fish breeders as often scars develop, which may be because the tissue here is very sensitive to extract.

2.12 Breeding Practices

Breeding of Indian major carp (IMC) and exotic carp is under-taken in captivity either through simulation of natural conditions in bundhs or through injection of hormones. Bundh breeding, which is mainly practiced in West Bengal, Madhya Pradesh, and Rajasthan, India, was introduced in the early 1950s and was initially considered the most dependable source of quality fish seed—once contributing 70% of the total seed produced in the state. Later, with the introduc-tion of hormone injection, several breeding devices were developed, such as breeding in hapa, Ciment cistern, community breeding pools, and breeding/spawning pools of masonry structure. The farmers in the modern hatcheries of Bengal use the hapa mainly for stocking of brooders before induction. Cloth hapa are also in use for breed-ing *Cyprinus carpio*. Some small farmers use hapa for raising a small number of seeds. Hapa are also used for stocking hatchlings before sale and for size-wise segregation of seeds.

Before 2000 most of the hatchery owners used to breed fish in breed-ing/spawning pools, but very recently the farmers have not been using this device and used to breed fish in an incubation pool, which is also used for hatching of fertilized eggs. Only 4–5% of farmers use commu-nity breeding pools for breeding large number of fish at a time. Breeding is followed by fertilization. For breeding and subsequent fertilization two procedures are generally adopted by the farmers. Hormonal induction operates differently with different fish species. For Indian carp, 6–8h after the second injection, the female and male are ready for spawning. But this is not the case for exotic carp, as they are less sexually respon-sive. Therefore, depending on the fish species and the sex, two devices are mainly used for breeding and fertilization, as described below.

2.12.1 Natural Breeding

In this procedure the fish are released into a breeding pool (now an incubation pool to save water) after hormonal induction. After sex play, which continues for 2–3h, both the female and male undergo spawning, which ensures fertilization.

2.12.2 Stripping

Stripping is generally practiced for exotic carp and catfish. For IMC it is done for hybridization and in acute cases of male brooder shortage. As it ensures complete spawning and increased fertilization, farmers practice it without considering the negative consequences, like damage to internal organs, stress to the brooders, mortality, etc. In the present-day hatcheries of Bengal stripping is also undertaken for common carp, *C. carpio*. This removes the hazards of natural breeding and also ensures better fertilization rate, requires less time, and lowers the risk in raising the seed. First, one person holds the tail region and another holds the head between the left hand and the body exerting pressure on the belly in an anteroposterior direction by placing the thumb on one side and the other fingers on the other side of the fish body, resulting in a steady, thread-like flow of eggs through the vent (Fig. 2.11A). Immediately after the complete release of eggs, the milt is stripped out from the male in the same tray (Fig. 2.11B), and this is followed by thorough mixing of eggs and milt with the help of a feather (Fig. 2.12). The fertilized eggs are then diluted with distilled water (Fig. 2.13). Diluted fertilized eggs are then released slowly in the incubation pool for further development. Stripping should be conducted with the onset of ovulation otherwise over-ripening may result in loss of viability and reduced fertilization. Though it differs according to species, in general loosening and associated ovulation occur within 5–7h of the second injection. For stripped bass and carp over-ripening does not create any problems if the fish is stripped, even after 14–18 days. For trout, turbot, and halibut, over-ripening leads to reduced fertilization and less viability of spawn if not stripped within 10 days of ovulation. For IMC and exotic carp, stripping is undertaken between 5 and 7h after the second injection. For pangus (*Pangasius sutchi*) stripping is done within 6–12h after the second injection. To achieve total success through stripping, periodic examination of brooders after the second injection is needed.

2.12.2.1 Stripping Time

For IMC and exotic carp, stripping is undertaken between 5 and 7h after the second injection. For pangasid stripping is done within 6–12h after the second injection, indicating more responsiveness of carp to GTH compared to catfish.

2.12.2.2 Environmental Influence on Stripping

Temperature is singled out as one of the crucial factors in controlling stripping; stripping time is enhanced at 30°C and above. Again, it is enhanced by 15–30min with the onset of sudden rain.

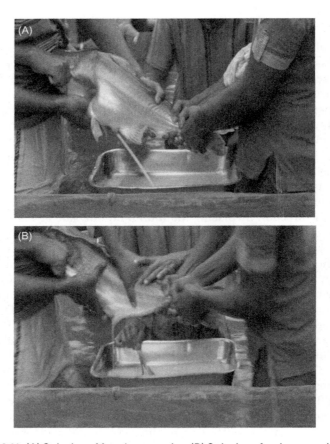

Figure 2.11 (A) Stripping of female pangasius. (B) Stripping of male pangasius.

Figure 2.12 Mixing of egg and milt of pacu by feather.

Figure 2.13 Dilution of fertilized eggs after stripping with distilled water.

2.12.2.3 Identification of Stripping Time

After the second injection the farmers keep the brood fish in hapa. Just before stripping the brood fish, particularly the females rub their body against the wall of the hapas which indicates the fish is ready for stripping. After the second injection the farmers maintain the brood fish in hapa until stripping. The females indicate the approach of stripping time by rubbing their body against the wall of the hapa.

2.12.2.4 Male and Female Ratio in Stripping

The success of stripping with increased fertilization rate mainly depends on the milt concentration of the male. If the male is mature enough and its milt concentration is optimum, then one male can fertilize the eggs of two or three females. After the second injection the farmers maintain the brood fish in hapa until stripping. The success of stripping with increased fertilization rate mainly depends on the milt concentration of the male. For trout 1 mL milt can fertilize 10,000 eggs, but if the amount of milt is more, then there is a chance of blocking the micropile of the egg so that fertilization will be impaired. It was observed that three to four drops concentrated milt can fertilize the eggs of one female.

2.12.2.5 Stripping Types

The mode of stripping is of two types:
1. *Wet stripping*: This involves stripping of eggs and sperm in a steel tray, half filled with water, with the understanding that water

facilitates easy approach of sperm towards egg to enhance fertilization. The entire procedure lasts for 10–15 s.

2. *Dry stripping*: After realizing the shortfalls of the wet method, the dry method was developed in the 19th century. In this process eggs and milt are stripped simultaneously in a steel tray (Fig. 2.11) and then eggs and milt are mixed thoroughly with a feather (Fig. 2.12). After through mixing, which lasts for 10–15 s, distilled water is added to the mixture (Fig. 2.13) which ensures complete fertilization.

2.12.3 Multiple Breeding

Multiple breeding of carp is practiced in about 80% of the hatcheries investigated. For Chinese carp, the breeding frequency is 3–4 times and for IMC it is 2–3 times, at a gap of 45 days during the breeding season. The quantity of eggs is maximum during the second phase of breeding (8–10 L), which coincides with the peak monsoon months, whereas during the first breeding (the third week of March), it ranges between 3 and 4 L. During the third breeding program, the quality of egg is moderate (6 L). It indicates that the breeders are forced to breed, during the first breeding program, though they are yet to attain complete gonadal maturity, which coincides with the vernal equinox. Moreover, the higher environmental temperature (35–37°C) as well as scanty rainfall are not optimum for an ideal breeding environment. On the other hand, the quantity of eggs is reduced during the third breeding program because of the following factors:

1. Improper maturity instead of enhanced feeding;
2. Environmental attributes;
3. Physiological shock, etc.

It is reported that fecundity remains unaltered during repetitive breeding of Chinese carp, whereas it decreases in the case of IMC in the subsequent breeding program. It indicates that Chinese carp are more efficient in gonadal recrudescence in comparison to IMC. However, the farmers are unaware of the fact that survivality and growth rate of the seeds generally reduces while practicing the multiple breeding of carps. Also, repetitive hypothecation within a short breeding season has been reported to cause blindness, particularly in males.

2.12.4 Mixed Spawning and Indiscriminate Hybridization

Mixed spawning and indiscriminate hybridization are customary practice in the present-day hatcheries of India. The fish breeders conduct mixed spawning by allowing fishes belonging to different species and genera of Indian carp, and sometimes also exotic carp, to breed in the same breeding pool or community breeding pool at the same time for the sake of convenience and time, and as it requires less water. As is known, reproductive isolation among different species of Indian carp and even up to genera are more dependent on external factors like geographical, ecological, developmental, and/or behavioral than reproductive, which is evident from the capture of intraspecific, interspecific, and intergeneric hybrids from native habitat. Identical chromosome number (Manna, 1984), identical isozyme gene expression (Padhi and Bukhsk, 1989), and conserved nature of some genetic markers (Ghosh and Sen, 1987) indicate a close phylogenetic relationship among different groups of carp. In all cases of hybridization the F1 hybrids and even F2 hybrids are fertile and viable. Hybridization is more frequently observed in fish compare to other vertebrate groups. Synchronization of spawning time, external fertilization, genomic plasticity, and transparent, nonadhesive, nonfloating nature of unfertilized eggs ease hybridization among different species and even genera of carp in nature. Again, a time period is required for the early development (cleavage, gastrulation, organ development) as the incubation period (16–18 h depending on temperature) is the same not only for different Indian carp but also for exotic fish also.

Along with this the fish breeders of India are regularly introducing exotic fish without maintaining any code of practice of introduction. The fish breeders, during lean periods, used to visit different European and even African countries to explore the possibility of importing alien species to India. Some of the already-introduced species are *Pangasius sutchi*, *Piaractus brachypomus*, *Clarius gariepinus*, *Clarius macrocephalus*, and other species not identified yet.

2.13 Identification of Fertilized Egg

During spawning, for identification, some eggs are collected from the hapa in a Petri dish or in a small tray, by slowly lifting one of the bottom corners of the open end. The eggs are then examined and, if found to be unfertilized, the experiment is discarded. But if the eggs are fertilized, they are taken to the laboratory for further study. Like the fertilized eggs, the unfertilized ones also swell up in water, but the color of the unfertilized eggs becomes an opaque whitish, while the color of fertilized eggs is crystalline transparent and the eggs look like pearls. Fertilized eggs are semibuoyant and come up to the surface on slight agitation of the water.

One interesting observation was noticed during stripping that color of eggs varies according to species, such as the color of carp are whitish, while that of pangasids is yellowish, and that of piaractus is greenish.

2.14 Characteristics of Carp Eggs

After fertilization, the eggs pass through a series of developmental stages and it is understood that to get maximum benefit of induced breeding the fish breeders should be well aware of each developmental stage. Within 15 min of shedding, the egg swells considerably by absorbing water through vitelline membrane (the thin outer envelope of the egg which swells up after shedding in water). The eggs have a glassy bead-like appearance, with a reddish tinge in catla, pale reddish in roué, light/pale brownish in mrigal, and pale bluish in calbasu. Immediately, a pinhead-like structure appears within developing eggs (Fig. 2.14). The major carp eggs have a large perivitelline space between the vitelline membrane and the egg proper in the center of the egg. The yolk is oval in shape with an elongation at its posterior end and is without oil globules. The posteriorly elongated oval-shaped yolk sac is a characteristic feature in all Cyprinids (carp). The yolk of the egg is more or less round in shape in the early period of development. It then gradually takes an oval shape. Afterwards a constriction appears in the posterior half of the yolk on its ventral side. The part of the yolk posterior to the constriction then elongates in the form of a tube, and the whole yolk thus takes the shape of a posteriorly elongated oval shape. The size of eggs from the same species of different breeders varies a good deal. The size of the fully swollen eggs of the four IMC (catla, rohu, mrigal, and calbasu) varies within the range of 2.5–6.5 mm diameter.

In the course of the development, the embryo of carp (young fish developing within the egg membrane) within the fertilized egg first appears as a spot, which develops gradually and takes more or less a pea-shaped appearance within approximately 4–5 h. The embryo

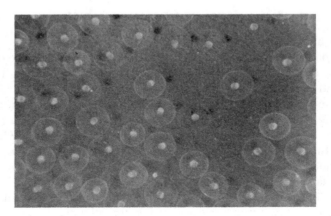

Figure 2.14 Pinhead-like developing embryo within water-hardened egg.

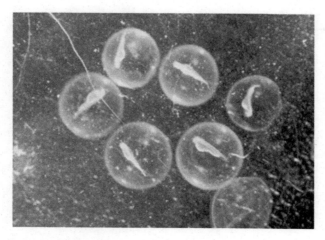

Figure 2.15 White thread or comma-shaped developing larvae with yolk sac within water-hardened eggs just before hatching.

then elongates and develops in the shape of a "comma" or fine white thread within water-hardened eggs, within about 7–8 h. This may be termed the "comma-shaped embryo" (Fig. 2.15).

It starts a twitching movement within approximately 9–12 h after fertilization. The rate of development of the embryo may vary, mainly according to the variation in the temperature of water. The higher the temperature the quicker is the development, and vice versa. In water temperatures ranging between 24°C and 31°C, the eggs usually hatch out within 14–20 h. During winter it takes 48–96 h for hatching and is evidenced during spawning and subsequent hatching of common carp eggs in winter. Besides temperature, water hardness and the presence of iron compounds in water impair hatching.

The newly hatched hatchling or spawn are fine thread-like with an oval yolk sac at the anteroventral side with a conspicuous head, two eyes, and a fine-thread behind (Fig. 2.16), which are absorbed by the growing spawn within 3 days and in the process the developing spawn attain the fry stage (Fig. 2.17). The first remarkable phenomenon in the development process is the "eyeing," which is initiated through the formation of pigmented retina. As a result the rudimentary eye becomes visible through the egg membrane, and immediately after elongated embryos start rapid movement, resulting in hatching and subsequent release of spawn or hatchlings to the external environment. The spawn still carry some of the yolk material and for this they are known as yolk spawn in the case of carp and alevin for trout and other fish. As the yolk is totally absorbed the spawn

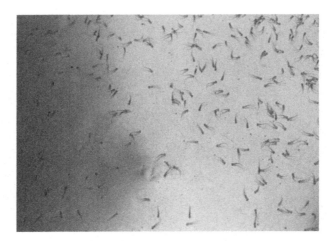

Figure 2.16 Spawn or hatchlings immediately after hatching (3 days old).

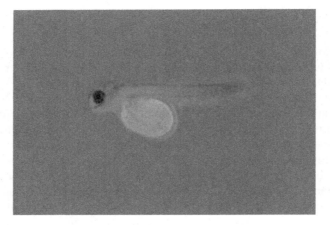

Figure 2.17 Hatched-out spawn with yolk sac.

starts its first feeding, and the spawn is considered vulnerable at this stage and immediately after hatching, eyeing and first feeding vary depending on species and temperature. In case of carp it takes 4 days and being aware of the same the fish breeders used to sell the spawn within 3 days. It is interesting to note that the size of the mouth opening may be small or large according to the size of the spawn or fry. Spawn and fry adjust their feeding habit according to the size of the mouth opening. In salmon fry, as the mouth opening is large they feed on artificial feed, but the spawn of carp, as the mouth opening is small, feed on small live plankton—a balanced diet needed for faster growth initially. Again the catfish fry are fond of eating chopped chironomous larvae. The fish breeders and farmers engaged in culture are often misguided by the agents and dealers of authorized/unauthorized feeds and fertilizers, and medicine companies as they are scientifically ignorant. To get a greater growth rate and to ameliorate the physicochemical conditions of water, fish farmers and breeders use unauthorized drugs, medicines, chemicals, feeds, fertilizers, oxygen enhancers, and many other products.

Probably the most preferred live food for early larvae is brine shrimp nauplii which needs decapsulation before application. As the time frame for decapsulation varies depending on the environment so proper care should be taken during its application. Nowadays several developments have been achieved towards production of dried and microencapsulated food. As the fry attain a size of 1 mm they are adapted to devour artificial food. During nursery rearing mustard oil cake and rice bran (1:1 by wt.) are applied as supplementary feed. Like many other animals, cannibalism is an important phenomenon which causes mass mortality due to cannibalistic wounds and subsequent infection by bacteria and fungus. A very common symptom the fish breeders encountered is the initial reddening of the anteroventral region followed by subsequent mass mortality of early spawn.

2.15 Development of Egg and Spawn With Special Reference to Indian Major Carp

Among cyprinids, three species of IMC, catla (*Catla catla* Hamilton), Rohu (*Labeo rohita* Hamilton), and Mrigal (*Cirrhinus mrigala* Hamilton) deserve higher economic importance due to customer preference. The original habitats of the three IMC are the rivers and backwaters of Northern India, Pakistan, and Burma. In addition, rohu inhabit the rivers of Central India and the south of Nepal (Terrain). The major carp have also been introduced into many other areas and countries. Though the IMC have not gained such widespread cosmopolitan acceptance as the common and Chinese carp, which are more adaptable and tolerant of a wider temperature range, they are nevertheless the most important cultured fish species in India, Pakistan, and Burma.

Under farm conditions, the IMC grow rapidly, reaching marketable size within 1 year in tropical and subtropical countries. The catla is the fastest growing species. In natural waters, it frequently attains 300–400 g in the first year, over 2 kg at the end of the second year, and 5–6 kg after 3 years. In fish farms, however, they can reach more than 1 kg during the first year. Rohu do not grow as rapidly as catla or the Chinese major carp, attaining around 500–1000 g in the first year under good farm conditions. The growth of mrigal is poorer than that of catla or rohu. As is the case for all cultured fish species, growth of farmed IMC is a function of stocking density, availability of natural and supplementary food, competition with other species in the same pond, and environmental conditions.

The food and feeding habits differ between the three IMC species. Rohu are "column" feeders, mainly collecting the zooplankton at all depths. They occasionally also feed on the bottom, utilizing organic detritus, remnants of aquatic plants, and mud rich in organic matter. Thus, when stocked together, catla and bighead carp can be competitors for the natural food supply. Rohu is mainly a bottom and column feeder, preferring plant material including decaying vegetation. The food preferences of rohu thus overlap with other cultured cyprinids less than those of the other IMC. Rohu have been used in Nepal to keep net cages free of fouling organisms such as sponges, bryozoans, and algae. Catla are a surface feeder and sustain on phytoplankton mainly. Mrigal have a narrower range of acceptable foods, being bottom-feeders living mainly on decaying vegetation. They are therefore similar in feeding habits to the Chinese mud carp.

Based on material provided by Dr. E. Woynarovich (Hungary) and Dr. V.R.P. Sinha (FAO Senior Aquaculturist, NACA, Bangkok, Thailand) the Indian carp do not breed naturally in ponds or other confined waters. They are all river spawners, having semifloating, nonadhesive eggs. The eggs have a large perivitelline space, and during swelling they increase their size by about 10 times (Fig. 2.16). Swollen, water-hardened eggs are rather delicate, and do not tolerate physical shocks well. This sensitivity to mechanical damage is a very important consideration when Indian carp eggs are to be incubated artificially.

The IMC normally attain sexual maturity in their second year. However, as with other cyprinids, age at sexual maturity is temperature- and light-dependent. It can be delayed by 1–2 years in the northern extremes of the fishes' range. The fecundity and breeding habits are similar in all the IMC, and they can be considered collectively. Though fecundity naturally depends on the nutrition and living conditions of each individual, generally females spawning for the second and third time produce the most ova, i.e., 100,000–200,000 eggs/1 kg of body weight. In nature, spawning coincides with the onset of the rainy season (depending on other climatic conditions). Details concerning the onset of development of ova in the ovary are still lacking. However, investigations have shown that in the northern hemisphere vitellogenesis starts in February and March. It ends after 1 or 2 months, and the ova are ready for final maturation and ovulation between April and June.

As the observations of the Brazilian scientist Rudolph von Ihering demonstrated in 1936–37, most river-spawning species of fish attain full ripeness of the ovary and become ready for induced propagation in confined, standing waters (ponds, lakes, etc.). The IMC are no exceptions to this rule and do not need any special treatment before hypophysation. The readiness for propagation is a natural requirement: the development of the oocytes must be finished in good time before the breeding season so that spawning can occur as soon as the environmental conditions become favorable.

Under natural conditions the IMC spawn in groups once a year. Thus at any one time all the eggs in the ovary are in approximately the same stage of development. In cultured conditions, they can be induced to breed twice within the same spawning season, the second occasion following about 2 months after the first. However, it is not yet known whether the second batch of eggs is the result of a new vitellogenesis or merely the remainder of the first batch which did not undergo final maturation and ovulation the first time. After natural spawning in selected stretches of the river, the semifloating eggs drift downstream. To survive and develop successfully, the eggs and larvae must reach inundated areas where the postlarvae and fry can find sufficient availability of food and remain safe from their enemies.

For centuries, before artificial propagation techniques were developed (in 1957), the eggs and larvae of IMC were collected from rivers using large funnel-shaped, fine-meshed nets, in much the same way as for Chinese carp. The collecting nets were continuously watched both day and night, and captured eggs and larvae were transferred to the safety of specially made small dug-out pools on the banks. The larvae collected thus were a mixture of all the river-spawning carps and sometimes other uneconomic species of fish. After a few days the number of larvae was estimated using small cups, and the fish were transported to fish farms for culture up to market size.

As fish seed is the major input for fish farming, there is always a growing demand for quality fish seed. But fresh and specific seed of a particular fish are not available everywhere due to lack of knowledge of identifying fish seed and profit-making seed production in the farming sector or in seed-producing industry. For this reason it is necessary to identify the fish seed before releasing them in the pond or any other farming sector for production and at the same time to prohibit inbreeding and other genetic consequences like genetic introgression and genetic drift, a common phenomenon in present-day hatcheries. Although it is too difficult to separate the different fish eggs or spawn from a mixture, the knowledge of developmental biology and morphological taxonomy will help in identifying the fish at an early stage. The fish breeders and farmers should be apprised of the same.

2.16 Larval Rearing

The fish breeders used to sell most of the seed produced, but the large fish breeders used to maintain a small percentage for culture; however, small and medium categories cannot afford to culture up to brood stage as it involves a minimum of 2 years. It should be kept in mind that at the initial stage, the spawn and the fry mainly depend on phytoplankton (e.g., chlorella, volvox, spirogyra, anabena, nostoc, vallisneria, dunaliela) and zooplankton such as rotifers (*Brachionus* spp.), copepod, flagellates (e.g., Isochrysis), brine shrimp, *Artemiasalina*, nauplii, and cladocerans like Moina and Daphnia.

References

Ghosh, D., Sen, S., 1987. Ecological history of Calcutta's wetland conservation. Environ. Conserv 14 (3), 219–226.

Manna, G.K., 1984. Progress in fish cytogenetics. Nucleus 27, 203–231.

Padhi, B.K., Bukhsk, A.R.K., 1989. Lactate dehydrogenase in the grass carp, Ctenopharyngodan idella (Pisces). Curr. Sci 58 (18), 1041–1043.

3

HYBRIDIZATION

CHAPTER OUTLINE

3.1 Natural Hybridization

3.1.1 In Bundhs

Hybridization may occur due to the congregation of too many species in the spawning ground, and due to limited space there is every possibility of ova of one species being fertilized by the sperm of another species of genetically flexible Indian major carp (Zhang and Reddy, 1991). During study of bundh breeding we frequently encounter natural hybrids. Hybrids, locally known as birbals, result from the (natural) cross-breeding of rohu and mrigal and are frequently observed in the bundhs of Bankura and Midnapore districts of West Bengal, India.

3.1.2 In Natural Habitat

Hybridization in nature among the three species of Indian carp, though not very common, is evident from the capture of hybrids from the wild on several occasions. These hybrids are found to bear intermediate characters of the two hybridizing species. The conditions leading to hybridization of fish in nature are not known, but synchronous breeding and spawning time, transparent, floating, and non-adhesive nature of the eggs of carp facilitate interspecific and even intergeneric hybridization, with consequent development of viable and fertile offspring.

Induced Fish Breeding. DOI: http://dx.doi.org/10.1016/B978-0-12-801774-6.00003-1

3.2 Artificial Hybridization

Artificial hybridization is mainly conducted through stripping as it ensures maximum fertilization. Experimental hybridization of Indian carp was first initiated in 1958 through stripping (Chaudhuri, 1959, 1961). Subsequently attempts were made to hybridize among the different species of Chinese carp and also between Indian × Chinese carp (Alikunhi et al., 1963) (Figs. 3.1 and 3.2).

Figure 3.1 Hybrid of *Labeo rohita* (♀) × *Labeo calbasu* ♂.

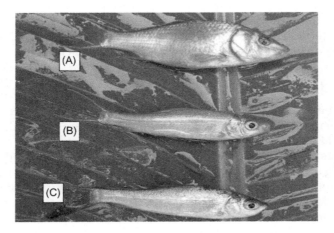

Figure 3.2 (A) *Cyprinus carpio* (male) × (C) *Labeo rohita* (female) = Hybrid (B).

3.2.1 Intraspecific/Interspecific and Intergeneric Hybridization

As has been recorded, initially six interspecific and 13 intergeneric hybrids have been developed through hybridization among the four species of Indian major carp (IMC), namely *Labeo rohita* (rohu), *Labeo calbasu* (calbasu), *Labeo bata* (bata), and *Labeo gonius* (gonius) in the following combinations. Interspecific and intergeneric hybrids among carps and also with common carp (*Cyprinus carpio* var. *Communis*, otherwise known as mirrorcarp) are found to possess useful cultural traits from an aquaculture point of view and are listed in Tables 3.1 and 3.2.

3.2.2 Outcrossing of Intergeneric Hybrids

Mature mrigal–calbasu hybrid female when crossed to male *Catla catla*, *L. calbasu*, and *Cirrhina mrigala*, produced the following hybrids:

Male Parent Species	Female Hybrid	Hybrid
1. *Catla catla*	Mrigal–calbasu	Catla–mrigal–calbasu
2. *Labeo calbasu*	Mrigal–calbasu	Calbasu–mrigal–calbasu
3. *Cirrhina mrigala*	Mrigal–calbasu	Mrigal–mrigal–calbasu

A good percentage of eggs hatched and of the hybrid fry that survived when reared, some males matured in 1 year.

Table 3.1 List of Interspecific Hybrids Between Indian Major Carp

Female	Male	Hybrid	Comments
1. *L. rohita* × *L. calbasu*		(rohu–kalbasu)	Hybrid grows faster than Kalbasu and fertile
2. *L. calbasu* × *L. rohita*		(kalbasu–roué)	Hybrid grows faster than Kalbasu and fertile
3. *L. rohita* × *L. bata*		(roué–bata)	Hybrids exhibited poor hatching and slow growth rate
4. *L. calbasu* × *L. bata*		(kalbasu–bata)	Hybrids exhibited poor hatching and slow growth rates
5. *L. gonus* × *L. calbasu*		(gonius–kalbasu)	Hybrids exhibited poor hatching and slow growth rates
6. *L. fimbriatus* × *L. rohita*		(fimbriatus–rohu)	Hybrids grow faster than fimbriatus and have better feeding efficiency

Table 3.2 List of Intergeneric Hybrids Among Indian Major Carp

Female	Male	Hybrid	Comments
1. *C. catla* × *L. rohita*		(catla–rohu)	Growth rate of hybrids variable, grow better than both the parents (Varghese and Sukumaran, 1971), grow slower than both the parents (Konda Reddy and Verghese, 1980), fertile hybrids
2. *L. rohita* × *C. catla*		(rohu–catla)	Hybrids grow faster than rohu. The meat yield in hybrids is higher than in parents. The hybrids are fertile
3. *C. catla* × *C. mrigala*		(catla–mrigal)	Hybrids grow slower than the parents and are fertile
4. *C. mrigala* × *C. catla*		(mrigal–catla)	Hybrids grow better than either parents and are fertile
5. *L. rohita* × *C. mrigala*		(rohu–mrigal)	Growth rate of hybrids variable, grow better than the parent species (Ibrahim, 1977), grow slower than both the parents (Basavaraju and Varghese, 1980b). The hybrids are fertile
6. *C. mrigala* × *L. rohita*		(mrigal–rohu)	Hybrids grow slower than both the parents and are fertile
7. *C. catla* × *L. calbasu*		(catla–kalbasu)	Hybrids grow faster than kalbasu and they have broader feeding spectrum and adaptability
8. *L. calbasu* × *C. catla*		(kalbasu–catla)	Hybrids exhibited low viability, need further study with regard to survival and performance
9. *L. calbasu* × *C. mrigala*		(kalbasu–mrigal)	Fertilization in the hybrid cross was normal (80–90%) with 60% hatching rate. The hybrids are fertile
10. *L. rohita* × *C. reba*		(rohu–reba)	Hybrids exhibit low viability
11. *L. calbasu* × *C. reba*		(kalbasu–reba)	Hybrids exhibit low viability
12. *C. catla* × *L. fimbriatus*		(catla–fimbriatus)	Hybrids grow much faster than *L. fimbriatus*. The meat yield is higher than the parents
13. *L. fimbriatus* × *C. catla*		(fimbriatus–catla)	Hybrids grow much faster than *L. fimbriatus*. The meat yield is higher than the parents
14. Rohu (♂) × mrigal (♀)		(Rni–Mirka)	Intermediate character between parents. Viable and fertile

3.2.3 Hybridization of Chinese Carp

Hybridization among *Aristichthys nobilis* (big head), *Hypophthalmichthys molitrix* (silver carp), and *Ctenopharyngodon idella* (grass carp) results in impaired hatching and survival of hatchlings up to 1 week only.

3.2.4 Hybridization of Indian Carp With Chinese Carp

Hybrids produced from crosses of Indian carp with different species of Chinese carp did not survive beyond 1 week:

3.2.5 Assessment of Unscientific and Indiscriminate Hybridization

1. Catla–rohu hybrid

    About 60% of the eggs undergo postfertilization development and usually hatching takes place after 16 h. Mortality of hatchlings is high, and the growth rate is faster than *L. rohita* fry of the same mother. The hybrid bears intermediate characters and is considered as the only hybrid which bears some cultural traits like head smaller than *C. catla* and body wider than *L. rohita*. The growth rate is faster than rohu but slower than catla. However, it has the advantage over catla in having a smaller head. The quantity of flesh is greater in the hybrid than that of catla. Hybrids attain maturity in 2.5–3.0 years. There are reports of poorly developed testes and development of hermaphrodite hybrids. Fecundity, however, was found to be less than that of the parents.

2. Catla–calbasu

    Initial development is abnormal and only 20% of the eggs hatched out and a small percentage survived. Growth and survival of hybrids were satisfactory and mature within the stipulated time.

3. Catla–mrigal

    Female traits are more inherited when reciprocal hybrids are compared, and it was noted that hybrids of catla females and mrigal males exhibited better cultural traits. Fertilization and hatching rates were much less, while very few reach maturity stage. The initial growth rate was satisfactory. Hybridization of rohu–calbasu and calbasu–rohu yielded 94% fertilization. The hybrids exhibits extreme variability and phenotypically resemble male parents. Reciprocal hybrids registered superior traits compared to slow-growing maternal *L. calbasu*.

    Embryonic development of all three hybrids of bata–rohu, bata–calbasu, and calbasu–gonius was normal though hatching percentage was poor. Since *L. bata* is a minor carp with a slower growth rate, the hybrids of bata–rohu and bata–calbasu were not profitable in regard to cultural traits, although the bata–calbasu hybrid was useful at the initial stage of development. The rohu–calbasu hybrid, on the contrary, attained maturity in 2 years. Hybrids bred normally in the pool, the fertilization rate was high, but the hybrids bore varying degrees of characters, intermediate between *L. rohita* and *L. calbasu* (Chaudhuri, 1960; Government of India, 1961).

3.2.5.1 Hybridization of Indian Carp With Common Carp

Stripped eggs of *L. rohita* and *C. catla* were fertilized easily by the milt of *C. carpio* (Alikunhi and Chaudhuri, 1959). The IMC eggs, through the process of water hardening, swell to the size of 4–6 mm in diameter, are nonadhesive and demersal, with an incubation period of 15–18 h at a temperature range of 26–31°C. In contrast, eggs of common carp are very small with limited perivitelline space, adhesive, with an incubation period of more or less 48 h depending on the temperature range. Both fertilization and mortality rates are high and larval death happens at hatching or immediately after hatching. All the other characters are intermediate except for an elongated dorsal fin, while adults did not mature. Reciprocal crossing was also tried. When the eggs of *C. carpio* were repeatedly fertilized by the milt of *L. rohita*, the fertilized eggs did not hatch. An out-cross with the female hybrid of rohu–calbasu and male *C. carpio* was successfully carried out. The hybrid was viable, fertile, and attained maturity within 2 years. In some farms of Bengal, India, when hybridization was attempted between scale carp (*C. carpio* var. *communis*) and rohu as well as catla, the hybrids in both cases were more or less devoid of scale. The skin is naked with patches of scale in a scattered fashion (rohu–cyprinus cross), while a linear row of scale is present along the mid-dorsal line (catla–cyprinus cross). They grow out normally, are fertile, and attain maturity within 2 years.

3.2.5.2 Experiments on Fish Hybridization in Other Asiatic Countries

Tang (1965) has reported successful hybridization of Chinese big head and silver carp. Details of morphological characters, growth rate, or behavior are not available. In 1966, during fish breeding experiments, attempts were made to cross medium-sized carp, *L. pangusia* with *L. rohita* and *C. carpio*. In both the cases stripped eggs of induced *L. pangusia* female were fertilized by sperm of *rohita*. Similarly another lot of eggs was fertilized by common carp sperm. The cleavage and early development, as well as embryo formation, were normal, without any observable deformity. But the development stopped suddenly at a later stage and the eggs did not hatch.

In Malacca Experimental Station some interesting hybridization work was carried out on *Tilapia* spp. Hickling (1960), through a hybridization experiment with *Tilapia mossambica*, demonstrated that when an African male was crossed with a Malayan female the hybrids were all males. In Indonesia several colored varieties and strains of common carp have been produced by selective breeding.

In Japan, hybridization led to the development of several varieties of gold fish (*Carassius auratus*) and common carp (*C. carpio*) through sustained selective breeding programs. Interspecific and intergeneric hybridization were mainly carried out on the cyprinid and cobitid fishes. Series of experiments on hybridization of the cyprinid fishes belonging to the subfamily, Gobiobotiini, have been conducted with a view to elucidate, based on morphological studies, whether or not the success of hybrid development is correlated with the supposed degree of phylogenetic relationship based on morphological studies (Suzuki, 1961, 1962, 1963).

Hybridization in USSR and Europe has been carried out as described here:

1. Domesticated common carp × wild common carp (in the former USSR and Europe)

   ↓↓

   Contamination in both and deterioration of economically important traits

2. *Acipenser sturio* (giant sturgeon or beluga) × *Acipenser ruthenus* (starlet)

   ↓↓

   F1-besters (possess good qualities of both parents)

   ↓↓

   Released into nature and causes considerable contamination of wild stock.

3.2.5.3 Evaluation of Hybrids

Keshavanath et al. (1980) evaluated the growth performance of hybrids produced out of a cross of *L. rohita* (female) × *C. catla* (male), and also the reciprocal hybrids along with both parent species in a community breeding pool as well as independently. These studies have shown that the growth of both hybrids was inferior to catla and almost equal to rohu. The growth difference was insignificant between the hybrids. However, when reared separately, the growth performance of the hybrids was better than rohu but inferior to catla.

The morphometric traits of rohu–catla and the reciprocal hybrids have been analyzed by different workers in comparison to the parent species. According to Konda Reddy and Varghese (1980) the rohu–catla and catla–rohu hybrids exhibit intermediate traits in respect of the morphometric ratios such as length of fish/length of head, length of head/snout length, length of head/width of mouth, and length of fish/length of dorsal fin. In other aspects like length of head/interorbital space, the hybrids tend more towards catla. Rohu–catla hybrids more closely resembled catla in the case of the ratios of length of head/diameter of eye, while the reciprocal hybrid catla–rohu tends more towards rohu. According to the authors, the taxonomic characters of these hybrids, with regard to meristic counts, have been observed to be similar. Almost all the hybrids, either interspecific or intergeneric, produced among IMC in general exhibited intermediate traits to their parental species.

3.2.5.4 Evaluation of Purposeful Hybridization

Experience with purposeful hybridization and back-crossing of F1 hybrids with parents is a regular practice in the seed production sectors, without further care of the hybrids this resulted in genetic introgression and contamination in the native gene pool. During investigation we have encountered a great deal of hybrids produced due to indiscriminate hybridization by fish breeders in India and with some scaleless hybrids. When scale carp is crossed to both rohu and catla, two phenotypically different hybrids were developed. Hybrids of scale carp and *C. catla* give rise to the development of lined hybrids (Fig. 3.3, having a single row scale along the mid-dorsal line), but when scale carp is crossed to *L. rohita*, it leads to the development of mirror carp, having scattered scale in some pockets. Works by Kirpichnikov and Balkashina (1935, 1936), Golovinskaya (1940), Kirpichnikov (1937), and Probst (1949) demonstrate that two pairs of autosomal linked dominant forms of alleles through epistatic interaction develop different types of scale pattern in common carp. The following carp genotypes SSNN, SsNN, SSNn, SsNn are possible.

Common carp with genotypes SSNN, SsNN, and ssNN are not viable as the N gene is lethal in the homozygous state and the embryos die at the hatching stage. Again, the gene exerts a reducing effect on development of some internal organs, when the carp is at heterozygous state (Nn; Golovinskaya, 1940; Kirpichnikov, 1945; Probst, 1953). This finding regarding hereditary pattern of basic

Figure 3.3 Intergeneric hybrid produced out of a cross between female *L. rohita* (A), male *catla catla* (B) and Hybrid (C).

scale arrangement has been confirmed by some recent workers. It is established now that S and N genes in European carp (*C. carpio* L) develop as a result of two independent mutations at two different loci. The combined effect of these two mutated genes developed leather carp, which is almost devoid of scale, leaving the skin naked. Both linear and leather forms of scale pattern are observed among the Japanese carp, *C. carpio haematopterus*. K.A. Golovinskaya claimed that heredity of scale patterns for these carp followed the inheritance pattern of the European subspecies (*C. carpio carpio*). Similar work has been conducted by Tran dinh-Trong (1967) and Kirpichnikov (1967) on mirror and linear carp from Vietnam (subspecies of *C. carpio fossicola*). Such characteristic hybrids were developed in the hatcheries of West Bengal, India, when mirror carp is crossed to *L. rohita* and *C. catla*. These cross-country experiments on development of mutated scale gene and their inheritance postulate that analogous scale genes have arisen in carp among the three subspecies which differ quite distinctly. These findings support Vavilov's law regarding homologous hereditary variation as expressed in fish (Vavilov, 1935).

As reported by Probst (1953) many internal organs are found to be poorly developed in both the linear and leather carp, which may be due to a defect in the development of mesenchyme during early development. It is established now that the dominant N gene developed out of a major deletion mutation resulting in a defect in the vital protein synthetic pathway. The defect in the development of scale as well as internal organs is due to deficiency of required protein resulting from the mutation of a single gene. This can be established by means of a serological test following a standard method. Following the same, Altuchov et al. (1966) and Pochil (1967) were successful in identifying the four mutated genes for the same by detecting erythrocyte antigens.

This mutated gene-related malformation includes defects in the formation of the vertebral column, varied degrees of reduction in gill cover, deformations of fins, etc., which are recorded for many families of carp. Similar abnormalities are encountered in the central regions of the Russian Federation, North Caucasus, and other regions of the former USSR. These deviations developed out of a decrease in the viability of heterozygotes with N gene (Nn) and mortality at homozygous state (NN). Varied structure of the swim bladder in carp results from pleiotropic effects of the "s" gene. When the "s" gene is present, the posterior chamber is typically smaller than the anterior chamber in scattered carp. However, in the presence of normal allele "S," the anterior chamber is longer (scaly carp). Carp with different scale genes may also be distinguished by means of serological methods.

Figure 3.4 Hybrid of rohu female and catla male (Nadeem).

These deviations in the development of morphological and anatomical abnormalities from one mutant to other vary from 0.1–0.2% to 20–30% and even more, the frequency being directly dependent on the degree of inbreeding and the environmental conditions under which the spawn and fry are reared. Considering a hereditary propensity to the malformation, some scientists are of the opinion that varied mutant types developed out of an interaction between the environment and genotype (Tatarko, 1961, 1966).

The mechanism of sex determination has yet to be confirmed in carp, due mainly to the large number of chromosomes in carp ($2n=104$). The diploid number of chromosomes of many cyprinids ranged between 50 and 52, which allows us to consider common carp as a natural (very ancient) tetraploid (Ohno et al., 1967). Hybridization attempts by scientists in different regions have yielded mixed results, which indicate that though there is potential for genetic improvement, considerable research needs to be done on a case-by-case basis before commercial application is considered. Nevertheless, one prediction that can be made is that successful hybridization between species that have different chromosome numbers, or strongly different karyotypes (numbers of acrocentric, metacentric etc., chromosomes), is very unlikely. Therefore detailed knowledge of the karyotypes of potentially hybridizing species is of significant importance for undertaking development programs through hybridization. One interesting observation noted during study was that hybrids inherit more maternal traits (characters) compared to male, and resemble females phenotypically (Figs. 3.4–3.6).

Figure 3.5 Hybrid of *Cyprinus carpio* var. communis (male) and *Labeo rohita* (female) (intergeneric).

Figure 3.6 Hybrid of *Cyprinus carpio* var. *communis* (male) and *Catla catla* (female) (intergeneric).

3.2.5.5 Evaluation of Hybrids in Respect of Meat Quality

Basavaraju and Varghese (1980a) have made some studies on organoleptic evaluation of the meat of rohu–mrigal and mrigal–rohu and compared it to that of the parental species, rohu and mrigal. The test was done for both raw and cooked meat.

The attributes tested were color gloss of skin, odor of flesh, color of flesh, and texture for raw fish, and color, texture, taste, and flavor for cooked meat. Analysis of variance (Krammer and Twigg, 1970; Basavaraju and Varghese 1980a) showed that the uncooked flesh of hybrids was found to be inferior in quality to that of the parental species, while no significant difference was found with regard to the cooked flesh.

3.2.5.6 Epistasis and Scale Pattern Inheritance in Carp During Hybridization

Some qualitative phenotypes in fish are controlled by two autosomal genes. When two genes control the production of a set of phenotypes, there is usually some sort of interaction and one gene influences the expression of the other. This means one gene alters the production of the phenotypes that are produced by the second gene. This gene interaction is called "epitasis." Most of the epistatic interactions were discovered in ornamental fish, but still several such phenomena have been discovered in economically important cultured fish like carp. Two such important epistatic inheritances are scale pattern inheritance in common carp and flesh color in chin hook salmon. As common carp are considered to be one of the most desired and economically important items for profitable fish culture throughout Asia and Europe, so the scale pattern is considered as an important item of study of qualitative inheritance in fish. It is known now that the scale pattern also helps to determine color pattern, and thus the value, of ornamental common carp (koi). Four phenotypic scale patterns, such as scaled carp (normal scale pattern), mirror, line, and leather carp are controlled by two dominant genes (S and N) through the genetic phenomenon known as "dominant epistasis."

The dominant "S" gene controls the basic scale pattern in common carp and the normal scale pattern both at homozygous (SS) and heterozygous (Ss) state. On the contrary, the recessive allele "s" gives rise to mirror phenotype (ss). Another dominant gene "N," through epistatic interactions with "S," modifies the phenotypes produced by the dominant "S" gene. Both these alleles (S and N) are two forms of a gene and are located in the same locus. Out of these two N is lethal at homozygous (NN) form. The lethal "N" allele, through epistatic interaction with dominant "S," produces three distinct phenotypes:

1. Dominant "N" allele, in homozygous state (NN), kills embryos (SS NN or Ss NN);

2. However, in heterozygous state (Nn), through epistatic interaction with "S" allele, it modifies scaly into line phenotype and mirror into leather phenotype (SS Nn or SsNn);

3. Recessive "n" has no effect on S and produces scaly phenotype (SS, nn or Ss, nn).

In heterozygous (Nn) state, the "N" allele changes the scaled phenotype into line phenotype and mirror phenotype to leather phenotype. The recessive "n" allele has no effect on the phenotypes produced by the dominant S gene. The five phenotypes (one is dead) and the genetics behind this phenotype are explained in Figs. 3.7 and 3.8 (Allendorf and Phelps 1980).

A set of qualitative phenotypes may be controlled by more than two genes. Body color in the Siamese fighting fish is an example of

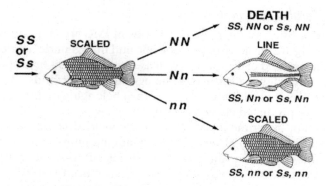

Figure 3.7 Scale pattern inheritance in common carp.

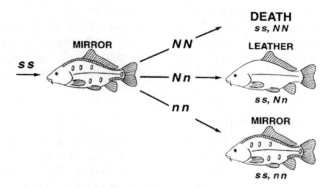

Figure 3.8 Scale pattern inheritance in mirror carp.

a set of phenotypes that is controlled by the epistatic interaction among four genes. Working with these phenotypes is far more complicated because of the number of genes involved. Fortunately, in food fish, qualitative phenotypes are not controlled by more than two genes, as detected by scientists still today.

3.2.5.7 Heterosis/Hybrid Vigor

The term heterosis (hybrid vigor) characterizes the increased ability of a hybrid compared to the parental forms. The term ability means the favorable changes in hybrid characters when compared to the abilities of parents in one or several characters. This phenomenon was discovered in the 18th century by Veldeiter during experiments on tobacco hybridization and was later confirmed by many selectionists working on hybridization of plants and animals.

3.2.5.8 Fertility of Hybrids

Interspecific and intergeneric hybrids of IMC are all fertile; again F2 offspring are fertile also. Interspecific and intergeneric hybrids like rohu–calbasu and catla–rohu as well as mrigal–calbasu bred successfully on inductions, again when the eggs of these hybrids are fertilized by the milt of all three species of carp belonging to three distinct genera they produced viable offspring.

In Japan, experiments with Cyprinids yielded some interesting results. Suzuki (1961) observed sterility among intergeneric hybrids, but at the same time intergeneric hybrids of *Gnathopogon elongates elongatus* and *Pseudorasbora parva* produced viable F2 progeny. It was noticed though the males of the intergeneric hybrid of *Pseudorasbora parvapumila* and *G. elongatus elongates* were fertile but the vast majority of the interspecific and intergeneric hybrids were either sterile males or neuters. It was reported that though the hybrid males bear normal testes, spermatogenesis was impaired. Observations made on the Indian carp showed that the majority of the interspecific and intergeneric hybrids, which grew to adult size, attained full maturity although poorly formed gonads were reported in some adult hybrids. Such discrimination in gonad development is not yet clear.

3.2.5.9 Monosex Hybrids

Fish culturists very often encounter the problem of overpopulation while dealing with prolific breeding fish like *Tilapia*. Since proper manipulation of stock is highly essential for obtaining higher production of fish in a pond, fish culturists need to find ways and means of preventing or controlling reproduction of these fish. The problem is more acute in countries where various species of *Tilapia* are cultured. *Tilapia mossambica* reproduces prolifically and soon overpopulates ponds, thus resulting in poor growth and poorer production of fish. Various methods, namely, monosex culture by segregating the males, has been tried with initial success but failed to meet the desired results. Inducing sterility or monosex production by cautious hybridization is an efficient method recently practiced for culturing the various *Tilapia* species. Hickling (1960) got all male offspring in *T. mossambica* from a cross of an African male (*T. hornorum zanzibarica*) with a Malayan female, which has considerable significance in *Tilapia* culture.

Since the hybrids are fertile, the fish culturist should be cautious enough to remove the parent stock from the spawning ponds to prevent their mating with the offspring. Further experiments in this line in hybridizing various *Tilapia* species have resulted in the production of 100 male progeny. Such hybrids were produced from the crosses of the following species:

Male Parent Species	Female Parent Species
1. *Tilapia hornorum*	*Tilapia mossambica*
2. *Tilapia hornorum*	*Tilapia nilotica*
3. *Tilapia nilotica*	*Tilapia hornorum*

The phenomena of andro- and gynogenesis could suitably be applied in *Tilapia* to produce monosex offspring for culture. Hybridization of Indian carp, both at interspecific and intergeneric levels, produce fertile offspring; the F2 hybrids are also fertile. Interspecific hybrids like rohu–calbasu and intergeneric hybrids, like mrigal–calbasu and catla–rohu, were both fertile. Outcrossing by fertilizing eggs with the milt of three species of carp belonging to three distinct genera were successful.

Suzuki (1961), while conducting experiments with Cyprinids, noticed varying results. He observed sterility among intergeneric hybrids, but the intergeneric hybrids of *G. elongates elongates* and *P. parva*, produce fertile F2 progeny. He also noticed that males of intergeneric hybrids of *P. parvapumila* and *G. elongates elongates* were fertile, but the vast majority of the interspecific and intergeneric hybrids were either sterile males or neuters. According to him, though these hybrid males had normally shaped testes, 21 allozyme loci in samples of wild-caught and hatchery-reared IMC from Bangladesh were analyzed. Bayesian model-based clustering analysis revealed the presence of four taxa, corresponding to the three known species along with a fourth unknown taxon present in two hatchery samples. Individual admixture coefficients showed that 24% of all hatchery-reared fish were hybrids, whereas a single hybrid was observed in the wild-caught samples. Only catla (*C. catla*) × rohu, *L. rohita* and mrigal (*Cirrhinus cirrhosus*) × rohu hybrids were observed, the vast majority of which were F1 hybrids, although five individuals represented putative backcrosses. Mitochondrial DNA analysis revealed that catla × rohu hybridization primarily involved catla males and rohu females, whereas mrigal × rogue hybrids primarily resulted from rohu males and mrigal females. Despite the high percentage of F1 hybrids in hatchery samples, reproductive barriers among species have so far precluded widespread introgression. Continued hybridization may eventually lead to a breakdown of species barriers, thereby compromising the genetic integrity of the species in the wild, and leading to production losses in aquaculture.

Genetic selection and hybridization of the cultivated species of fish of Asia have to be undertaken with a view to evolve strains which have quicker growth rates, higher fecundity, earlier maturity, better taste and nutritional qualities, more efficient food conversion, and that are more resistant to unfavorable environmental conditions, diseases, and parasitic infections. Progeny testing should follow every program of hybridization. This involves the estimation of breeding worth on the basis of the performance of the offspring. Since both fertility, and induced sterility in hybrids are of vital importance in fish culture, more stress should be given to the study of genetic causes of sterility and reduced fertility in hybrids. Prevention or control of reproduction for proper stock manipulation in the cultivated varieties of fish which readily reproduce in ponds is no less important than making the difficult-to-spawn fish spawn or improving the stock by genetic selection and hybridization. Since this aspect of research has received very little attention, not only in Asia but also in most other parts of the world, it is suggested that this particular aspect of fish culture should receive due attention of all fish culturists. The hazards of indiscriminate hybridization cannot be overstressed. Fish culturists should be very cautious in their selection and hybridization programs and ensure that worthless hybrids are not propagated.

References

Alikunhi, K.H., Chaudhuri, H., 1959. Preliminary observations on hybridization of the common carp (*Cyprinus carpio*) with Indian carps. Proc. Indian Sci. Congr. 3, 46.

Alikunhi, K.H., Sukumaran, K.K., Parameswaran, S., 1963. Induced spawning of Chinese grass carp, *Ctenopharyngodon idellus* (C & V) in ponds at Cuttack, India. Proc. Indo-Pacif. Fish Coun. 10 (2), 181–204.

Allendorf, F.W., Phelps, S.R., 1980. Loss of genetic variation in hatchery stock of cutthroat trout. Trans. Am. Fish. Soc. 109, 537–543.

Altuchov, Iu.P., Matveeva, A.N., Pusanova, B.M., 1966. Heat-resistance of muscular tissues and antigens properties of erythrocytes in some cultured carp varieties. Tsitologia 8 (1), 100–104. (In Russian).

Basavaraju, Y., Varghese, T.J., 1980a. A comparative study of the growth of rohu-mrigal and mrigal-rohu hybrids and their parental species. Mycore J. Agric. Aci. 14, 388–395.

Basavaraju, Y., Varghese, T.J., 1980b. Organoleptic evaluation of flesh of two major carp hybrids. Curr. Res 9, 47–48.

Chaudhuri, H., 1959. Experiments on hybridization of Indian carps. Proc. Indian Sci. Congr. 46 (4), 20–21.

Chaudhuri, H., 1960. Experiments on induced spawning of Indian carps with pituitary injections. Indian J. Fish. 7, 20–48.

Chaudhuri, H., 1961. Spawning and hybridization of Indian carps. Proc. Pacif. Sci. Congr. 10 (Abstract only).

Golovinskaya, K.A., 1940. Pleiotropic effect of scale genes in carp. Dokl. Akad. Nauk SSSR 28 (6), 533–536.

Government of India, 1961. Annual report of the Central Inland Fisheries Research Institute, Barrackpore, for the year 1960–61. Indian J. Fish. 8 (2), 526–574.

Hickling, C.F., 1960. The Malacca tilapia hybrids. J. Genet. 57 (1), 1–10.

Ibrahim, 1977. Contribution to hybridization among Indian Carps: studies on the biology and morphological features of some Indian Major Carpshybrids. Utkal University Ph.D.thesis.

Keshavanath, P., Varghese, T.J., Reddy, P.K., 1980. Preliminary studies on the growth of catla-rohu and rohu-catla hybrids. Mysore J. Agri. Sci 14, 401–407.

Kirpichnikov, V.S., 1937. Principal genes of scale in carp. Biol. Zh. 6 (3), 601–632.

Kirpichnikov, V.S., 1945. The influence of environmental conditions on viability, rate of growth and morphology of carp (of) different genotypes. Dokl. Akad, Nauk SSSR 47 (7), 521–525.

Kirpichnikov, V.S., 1967. Homologous genetic variability and evolution of wild carp. Genetika 3 (2), 34–47. (In Russian).

Kirpichnikov, V.S., Balkashina, E.J., 1935. Materials on genetics and selection of carp. 1. Zool. Zh. 14 (1), 45–78. (In Russian).

Konda Reddy, P., Varghese, T.J., 1980. Embryonic and larval development of the hybrid. Catla-rohu. Mysore J. Agri. Sci 14, 588–598.

Krammer, A., Twigg, 1970. Quality Control for Food Industry. The Avi. Publishing co. Inc, Westpost, CT.

Ohno, S., et al., 1967. Diploid-tetraploid relationship among old-world members of the fish family Cyprinidae. Chromosoma 23 (1), 9.

Pochil, L.J., 1967. The erythrocytes antigens of carp (*Cyprinus carpio* L.), grasscarp (*Ctenopharyngodon idella Vall.*) and crucian carp (*Carassius auratus gibelio* Bloch). Trudy Vsesoiuz. Nauchno-issled. Inst. Prud. Ryb. Khoz. 15, 278–283.

Probst, E., 1949. Vererbungsuntersuchungen beim Karpfens. Allg. Fischztg. 21, 436–443.

Probst, E., 1953. Die Beschuppung des Karpfens. Münch Beitr. Abwass.-Fisch. u. Flussbiol. 1, 150–227.

Suzuki, R., 1961. Sex and sterility of artificial intergeneric hybrids among bitterling (Cyprinid fishes). Bull. Jap. Soc. Sci. Fish. 27 (9), 831–834.

Suzuki, R., 1962. Hybridization experiments in Cyprinid fishes. I. *Gnathopogon elongatus* female *Pseudorasbora parva* male and the reciprocal. Bull. Jap. Soc. Sci. Fish. 28, 992–997.

Suzuki, R., 1963. Hybridization experiments in cyprinid fishes. III. Reciprocal crosses between *Pseudorasbora parva pumila* and *Gnothopogon elongatus elongatus*. Bull. Jap. Soc. Sci. Congr. 29, 421–423.

Tang, Y.A., 1965. Progress in the hormone spawning of pond fishes in Taiwan. Proc. Indo-Pacif. Fish. Coun. 11, 332–334.

Tatarko, K.I., 1961. The aberrations in structure of gill cover and fins in carp. Vop. Ikhtiol. 1 (3), 412–420.

Tatarko, K.I., 1966. Anomalies of the carp and their causes. Zool. Zh. 45 (12), 1826–1834. (In Russian).

Tran dinh-Trong, 1967. Data on the intraspecific variability, biology and distribution of the carps in North Vietnam (Democractic Republic of Vietnam). Genetika 3 (2), 48–60. In Russian.

Varghese, T.J., Sukumaran, K.K., 1971. Notes on hypophysation of catla-rohu hybrid Annual Day Souvenir. CIFE, Bombay.16–17

Vavilov, N.J., 1935. The law of homologous lines in hereditary variation. Teoret. Osnovy Selekts. Rast. 1, 75–150.

Zhang, S.M., Reddy, P.V.G.K., 1991. On the comparative karyomorphology of three Indian major carps, *Catla catla* (Ham.) *Labeo rohita* (Ham.) and *Cirrhinus mrigala* (Ham.). Aquaculture 97, 7–12.

4

SELECTIVE BREEDING

CHAPTER OUTLINE

Induced Fish Breeding. DOI: http://dx.doi.org/10.1016/B978-0-12-801774-6.00004-3

4.1 Genetic Improvement of Farmed Fish

The application of genetics has helped to enhance the production of some crop plants, dairy milk, and eggs. It has also helped in the increased production of some fish species. It is anticipated that further improvements in aquaculture productivity as a whole will be possible through the application of genetic principles and technology. Genetic upgradation in farmed fish can be done by (1) combining good genes and/or (2) incorporation of useful genes.

The procedure used for genetic upgradation includes:

1. Selective breeding;
2. Hybridization;
3. Chromosome manipulation;
4. Gene manipulation.

Selective breeding involves the selection of superior-performing individuals out of a population to be used in a successive breeding program with the objective of producing heterotic offspring for the next-generation breeding program. It encompasses the following steps:

1. Trait selection in a species;
2. Choice of breeding strategy;
3. Selection method;
4. Evaluation of selection response.

Selection entails choosing some individuals from the population to produce more offspring than others. These individuals are selected from the animals which reach sexual maturation, as only those which can reproduce can affect the future population. The aim of selection is to identify and select parents for the next generation the individuals whose progeny, as a group, have the highest possible additive genetic merit for the trait or traits in question. Selection does not create new genes, but rather changes gene frequencies. The frequencies of alleles with favorable effects on the phenotype under selection are increased, and the frequency of less favorable genes is decreased. If the purpose of selection is to improve a production trait, the first step is to measure or record this trait for all animals in the population, and then estimate the average and standard deviation. Selection is then practiced by selecting those animals which have the highest breeding values. The genetic worth of an individual brood stock can be judged by several methods such as (1) mass selection; (2) between-family selection; (3) within-family selection; and (4) combined selection. The contribution of offspring to the next generation is called the *fitness, adaptive value,* or *selective value* (Falconer and Mackay, 1996) and fitness is the "character" that natural selection selects for. Animals that have the highest fitness in the current environment will reproduce at a higher level and have a higher survival rate than less fit animals. The effect of natural selection over generations is to establish a population adapted to the new environmental condition. Thus, adaptability is a response of a population rather than of an individual. Natural selection is rather inefficient over short time periods. One of the reasons for its inefficiency is that it only utilizes individual or mass selection. Natural selection can be defined as the differential contribution of genetic variation to future generations (Aquadro et al., 2001) due to differential reproduction of some phenotypes/genotypes over others under prevailing environmental conditions at a given time (Futuyma, 1998). It is the driving force behind Darwinian evolution and can be subdivided into different types, depending on several factors of nature.

Artificial selection operates the selection process in a desired direction. However, at the same time the population will continue to be affected by natural selection, which may act in the same or in the opposite direction as artificial selection. Artificial selection may be performed based on the following criteria:

1. Animals with average performance are selected.
2. Extreme animals are selected and mated intersex.
3. Animals with good performance in one direction are selected and are practiced by selecting phenotypes around the mean and discarding extremes.

The aim of stabilizing selection is to standardize the population around an average and results in a fairly constant mean with somewhat reduced variance for the trait in question. In stabilizing selection some traits there may be an intermediate optimum which is dependent on the biology and life history of the organism, such as in fish an optimum fat content may be wanted in the fillet, while high and low fat percentages are less desirable. Diversifying selection extremes selected parents in each extreme group which were mated, offspring variance would be increased and eventually separate and distinct subpopulations could emerge. This method of selection is rarely used in animal breeding.

4.1.1 Directional Selection

The most common form of selection applied in agriculture and aquaculture breeding programs is *directional selection.* The aim is to improve traits of economic importance. The effect of directional selection for heritable traits is a change in gene frequency at the loci affecting the trait in the next generation. Assuming no change in environmental conditions the average phenotypic value of progenies of selected parents is increased. Directional selection tends to decrease variation *within* a population but may increase or decrease variation *among* populations. Positive selection is a type of directional selection that favors alleles that increase fitness of individuals. When directional selection eliminates unfavorable mutations, it is called purifying selection (also known as negative selection).

Diversifying (or disruptive) selection favors variety and benefits individuals with extreme phenotypes over intermediate. In this type of selection, the propagation of an allele never reaches fixation, and therefore it may occur when an allele is initially subject to positive selection, and then negative selection when the frequency becomes too high (Nielsen, 2005).

Multiple Trait Selection

A selection program which usually focus on several traits of economic importance and basically involves three methods (Hazel and Lush, 1942) such as (1) tandem selection, (2) cull simultaneously, and (3) selection index or total score.

Tandem Selection

The selection in which several traits are improved in succession, such as selection is carried out first for one trait, until a desired genetic level is reached, then selection for the second, third, and following traits are performed.

Cull Simultaneously

The second method is to cull simultaneously but independently for each trait, a selection strategy called independent culling levels. Then a level is set for each trait which represents the culling or selection level for each trait.

Selection Index or Total Score

The third method is to apply selection simultaneously to all traits providing appropriate economic weight, heritability, and phenotypic and genetic correlations between the traits.

Of the three, the tandem method is least efficient, but the selection index method is the most efficient in terms of selection response in the direction of the multiple trait breeding goal.

Indirect Selection

This method may be exploited in breeding programs when a trait is difficult or very expensive to measure or record, a correlated trait may be used instead, may be of importance, or may have no economic importance at all.

Marker traits include (1) disease resistance selection growth rate; (2) feed conversion efficiency; (3) absorption of nutrient; and (4) reduced fat deposition.

Disease Resistance Selection

As we know, growth is also correlated with disease resistance, or at least with survival selection for growth involves a positive correlated response in survival. Immunological and physiological parameters such as (1) lysozyme (Lund et al., 1995), (2) hemolytic activity (Røed et al., 1990), (3) cortisol (Refstie, 1982), (4) IGM (Lund et al., 1995), (5) antibody titer (Lund et al., 1995), and (6) plasma $\alpha 2$ antiplasmin activity have shown that genetic variation and genetic correlations related to survival but none of these genetic correlations exceeded ± 0.37, so the correlated response would not be expected to be particularly large.

4.1.2 Balanced Selection

Balanced selection, which helps to maintain an equilibrium point at which both alleles remain in the population, has several forms, including frequency-dependent selection and overdominance, which occurs when the heterozygote has the higher biological fitness, and therefore variability is maintained in the population (Nielsen, 2005).

4.1.2.1 Mass Selection

This is the simplest form of selection, where the best individuals are selected from a population on the basis of their own "*phenotypic value*" compared to the population mean. Generally the top 10–15% of individuals are selected and it does not require individual identification or the maintenance of pedigree records, hence it may be considered the least costly method. In principle, it can produce rapid improvement if the heritability of the trait(s) under selection is high. Under those circumstances, however, there is risk of inbreeding due to inadvertent selection of progeny from few parents producing the best offspring, especially if progeny groups are large. If controlled pair matings can be carried out, the results of Bentsen and Olesen (2002) can be used to formulate the design of the breeding program. These authors investigated the effect of number of parents selected and of number of progeny tested per pair for a range of population sizes and heritability values. They show that inbreeding rates can be kept as low as 1% per generation if a minimum of 50 pairs are mated and the number of progeny tested from each pair is standardized to 30–50 progeny. Note that although not requiring individual identification of the fish, the schemes suggested by these authors entail the conduct of pair matings, initial maintenance of the progeny of such pair matings in separate enclosures, and controlled contribution of each full sib family to the next generation at the time the fish are assigned to communal rearing.

4.1.2.2 Between-Family Selection

Of the several families only those exhibiting good phenotypic value are selected for between-family selection.

4.1.2.3 Within-Family Selection

In within-family selection each family is considered a subpopulation and an individual from a family is selected or culled based on their relation to their family mean. This method is more efficient than the above two methods. Within-family selection to improve growth at 16 weeks was undertaken on Nile tilapia (*Oreochromis niloticus*) from 1986 to 1996. Data from 12 generations onwards were analyzed using a single trait restricted maximum likelihood fitting an animal model. The heritability in the base population was estimated as 0.385. A genetic response for body weight was found with an expected mean increase in body weight per generation of 2.2 g or about 12.4%. A realized heritability estimate of 0.14 was obtained based on the regression of mean breeding values on cumulative selection differentials after 12 generations. The inbreeding coefficient was 6.3% after 12 generations with an average inbreeding of 0.525% per generation. The family rotational mating used to propagate the families was effective in keeping the inbreeding rate to a minimum even at high selection intensities. High selection intensities and the relatively high heritability for body weight at 16 weeks resulted in a substantial response using the within-family selection method.

The two most common selection systems used during early selection process are individual or mass selection and family or selection of clones. It is common in most breeding programs to use an individual/mass selection procedure as described by Skinner (1971). Hogarth (1971) showed that family selection is more effective and economical than individual selection at the primary stage of selection.

4.1.2.4 Combined Selection

Within-family selection and between-family selections can be combined in a single program where the best individuals of the best families can be bred together. The advantage of this method is that the additive genetic variation between and within the families can be exploited to have a better selection response. A selective breeding program has helped to improve the aquaculture.

4.1.3 Selective Breeding and Disease Resistance

Selection is considered as an efficient tool to develop disease resistance in fish in the near future and, and in some specific cases has already been proved to be an important means for control of fish diseases. Selection work often necessitates transportation of fish from one area to another and this often leads to introduction of diseases and parasites into new pond establishments and commercial waters previously absent. As we know, resistance to a specific disease is dependent on a relatively simple mutation involving one or a few genes. Generally gene mutation is accompanied by the synthesis of changed proteins and may result in protein incompatibility of parasite and host (Kirpichnikov et al., 1967). A selective breeding program to develop higher resistance to disease is a recent approach in fish culture and so the data available are scanty. Insufficient study of fish diseases, especially the infectious ones, and contradictory observations on immunological reactions in fish have also contributed to this. Until recently some authors considered that fish are not capable of producing antibodies, however recent findings indicate fish do produce antibodies but not as intensively as higher vertebrates. Immunobiological reactions of fish are undoubtedly affected to a great extent by water temperature; the higher the temperature, the more pronounced the reaction. At very low water temperature, the processes of antibody formation are almost imperceptible (Krantz et al., 1964; Vladimirov, 1966).

Bactericidal activity of organs and tissues has also been discovered in fish. It is probably dependent on the presence of such antimicrobial proteins as complement, lysozyme, and properdian (Lukyanenko and Mieserova, 1962; Lukyanenko, 1965; Bauer et al., 1965). At temperatures lower than 10°C the mechanism of excretion and bactercidal action of organs and tissues predominate over other immune mechanisms (Avetikyan, 1959). Phagocytosis, a mechanism of antibody formation, begins to function with an increase in temperature (Goncharov, 1963, 1967). Recent data on fish immunity provide selection of breeders with a theoretical background for the breeding of disease-resistant fish strains.

4.1.3.1 Selective Breeding of Disease-Resistant Fish

When a disease infects fish, they are not all affected to the same extent, instead some fish (may be larger or smaller) are not affected. Schäperclaus (1953) cites this example: In the spring of 1950 a small pond in Peitz fish farm was stocked with 665 two-year-old carp which were kept there until early June. In June, an outbreak of infectious dropsy occurred. When the pond was fished out, it was found that 4% of the fish were either dead or doomed to die; 54% of the fish had external ulcers typical of the disease; and 42% showed no symptoms of the disease. Even progeny of a single pair of spawners reared under the same environmental conditions did not all possess the same resistance. Individual specimens may serve as starting points for selection work. Fish belonging to different strains and forms of any species differ in their resistance to disease. Wild carp (*Cyprinus carpio*) and especially its Amur subspecies (*C. carpio haematopterus*) are known for their higher resistance to infectious dropsy than cultured carp (Karpenko and Sventycki, 1961). Amur wild carp × cultured carp hybrids also possess high resistance. The replacement of cultured carp with hybrid forms has happened in recent years in a number of Ukrainian fish.

The same was observed in relation to another widespread infectious disease of carp, the air-bladder disease; it is widespread, not only in the USSR but in the other European countries.

Arshaniza (1966) carried out experiments on joint rearing of the cultured carp and the hybrid of the fourth generation of cultured carp × Amur wild carp. The results showed that the incidence of air-bladder disease amounted to 78% in the former, while it occurred in only about 30% of the latter. Symptoms of illness were more pronounced in the cultured carp than in the hybrid. Considering these examples, we may assume that work on the development of disease-resistant strains of fish may be started not only with the selection of specimens possessing higher individual resistance (fish that remained healthy within a stock affected by the disease), but also with the development of breed groups of hybrid origin, distinguishable by higher resistance to a particular disease. Naturally, such selection work should be carried out at those fish farms which are always under the effect of the disease. Otherwise the increased disease resistance of the selected stock would easily be lost. As stated earlier, data are scant on the selection of disease-resistant stocks of fish, and many of the studies undertaken on the problem have either not been carried through or are still in progress. Only three examples of such studies have been carried to conclusion, two of which concern infectious dropsy.

For a long time dropsy was considered a bacterial disease, *Aeromonas punctata* being its pathogenic agent (Schäperclaus, 1930, 1954). Subsequently, many scientists have questioned the bacterial nature of the disease and have come to the conclusion that dropsy is caused by a virus. Direct evidence of the viral nature of dropsy was obtained several years ago by Yugoslavian scientists (Tomasec et al., 1964). They succeeded in growing the causative agent of dropsy on artificially cultured tissue of carp kidney, and infecting healthy fish with it. The second International Symposium on Fish Diseases held in Munich in 1965, recognized a virus as the primary pathogenic agent of dropsy, while *A. punctata* and other saprophytic bacteria complicate the course of the disease.

4.1.3.2 Breeding of a Dropsy-Resistant Strain of Carp

Attempts to create a stock of dropsy-resistant carp were made for the first time by Schäperclaus (1953). Work on selection of dropsy-resistant carp was started by him in 1935. Carp not affected by the disease, in ponds where dropsy outbreak had occurred, were chosen for selective breeding. For two successive years the carp were exposed to large amounts of *A. punctata* with a view to artificially strengthening their natural immunity. Fish having even the slightest symptoms of disease were eliminated. The offspring of both selected and nonselected fish were compared for dropsy resistance; the results showed that of the former only 2% contracted dropsy, but of the latter 30% did. No dropsy was recorded among the progeny of the selected spawners. From 1939 to 1943 the mortality rate for the ponds (65 ponds with a total area of 180 hectares) where the progeny from selected spawners was reared, averaged 11.5%, its range varying between 0% and 37%. The average mortality rate for the control ponds (76 ponds with a total area of 474 hectares) was 57%, its range being 4–96%. A decrease in mortality was observed in subsequent years as well. From these observations Schäperclaus concluded that resistance to dropsy is inheritable and it is possible by means of correct selection to obtain stocks possessing relative immunity to this disease.

However, it is possible to obtain this effect only where a selected stock is not mixed with any other and where no infection is brought in from outside, either with the water or with the fish delivered to the station. Experience over many years has shown that infection brought into a specially selected stock manifests, as a rule, in an acute form. But such complete isolation of the selected stock cannot be maintained in the carp farms situated within the region of the natural habitat of *C. carpio*. Infection is continually brought in either by wild fish penetrating the fish farm, or through water exchange. or possibly by some other unknown agent. As suggested by Kirpichnikov et al. (1967) fish farmers should select spawners from breed groups that are most resistant to dropsy and should further increase their resistance by means of a controlled selection program. To start with Kirpichnikov first initiated such a program at fish farms of the Northern Caucasus with the aim to compare the resistance of different breed group of carp to dropsy. The following groups of carp were tested, singly as well as in mixed groups.

1. Ropsha carp (hybrid Amur wild carp× mirror carp, F_4 and F_5);
2. Selected Ukrainian carp;
3. Local scaley carp and low-bred mirror carp of unknown origin;
4. Wild carp of the Don River.

The aforesaid trials for development of resistance to dropsy were carried out at one of the fish farms and the larvae of an individual experiment and also the different breed groups were combined and reared for 3 years (1963–1965). Individual groups were marked by fin clipping, but differential growth of fin yielded errors toward differentiation of the different breed groups. Experimental results indicate that different breed groups exhibit differential resistance to dropsy, but the most resistant group was Ropsha carp, local carp and wild carp and increased resistance may be assigned to the wild origin of the said carp. Besides high disease resistance, the Ropsha carp exhibited higher viability, and were therefore selected for a next-generation breeding program at these fish farms. Afterwards, 3-year-old fish of all groups were considered as brood stock for the second-generation breeding program, and were subjected to intensive selection for developing further resistance to dropsy. This experimental work promises great economic possibilities after its completion.

4.1.3.3 Breeding of Disease–Resistant Trout

In the early 1950s work was initiated for the development of two separate disease-resistant (furunculosis and ulcer) trout stocks, brown trout (*Salmo trutta m. fario*) and the brook trout (*Salvelinus fontinalis*) (Wolf, 1954). The primary objective was to test the susceptibility of different fish populations to the diseases. To evaluate the same, fish collected from the ponds of diverse geographical territories were stocked in a particular pond of the experimental farm. Then some disease-affected fish were released into these experimental ponds. Diseases broke out in all the experimental ponds, causing mortality. The mortality rate of the brown trout varied from 35.7% to 65.8%, while for brook trout the range was 5.3–99.6%. The results indicate that as the brown trout are less susceptible or higher resistant to disease so they will be considered for further selection work (Elinger, 1964).

Diverse geographical strains of brown trout and brook trout were intensively exposed to *Aeromonas salmonicida*, and the less-resistant individuals were removed by mass selection. Survivors were selected as brood fish for the next generation, which was similarly exposed to the disease. Hybridization through crossing more promising strains in some instances improved resistance. Resistant strains were compared with regular hatchery stock under actual hatchery conditions. The results obtained thus far indicate that resistance to furunculosis is being achieved. The work is in progress in some leading microbial research laboratories and is expected to yield good results with the development of a highly resistant strain.

4.1.4 Effect of Selection Work on the Distribution of Parasites and Disease of Fishes

For carrying out selective breeding it is necessary to collect fish belonging to different populations, forms, subspecies, or species from diverse geographical territories and their successive rearing in experimental fish farms. The negative aspect of this collection procedure is the transfer of disease into the experimental farm, with either eggs/spawn/fry/fingerlings or adults. Infection is negligible when eggs are chosen as the initial material for selection. One such example is the transfer of pathogenic agent of whirling disease, a most troublesome trout disease (*Myxosoma cerebralis*) with trout eggs. It is assumed that the spores of the disease organism are introduced into the water bodies of new fish farms, however no such corroboration has yet been established. Efforts in obtaining carp eggs under hatchery conditions, with a view to propagate them artificially, have shown that healthy progeny may be obtained from the eggs of diseased fish. It has been demonstrated that such widespread diseases of carp as infectious dropsy and air-bladder disease are not transmitted germinatively. However, there exists the possibility of introduction of parasites infesting the eggs into new bodies of water and fish farms. Fortunately, such parasites are few. Only one species of Coelenterate, infesting acipenserid eggs (*Polypodium hydriforme*) has been recorded in the former USSR. The danger of bringing it along with eggs is not excluded, and every precaution should be taken while transporting eggs for fish culture purposes.

The danger of possible introduction of pathogens while transferring fry or older fish is more real. Soviet ichthyo-pathologists have collected a good deal of data on this. The following is one example. From 1958 to 1962, large numbers of fry of the grass carp *Ctenopharyngodon idella*, caught in the rivers of China, were transferred to the USSR. In spite of all the precautions taken, more than 20 species of parasite were introduced with these fish. Many of these parasites cause severe epidemics, not only among phytophagous fish, but among other cultured and commercial fish as well. Thus the cestode *Bothriocephalus gowkongensis* came to be introduced into the fish farms of the European part of the USSR and Central Asia; then it infected the carp and many other fish, causing high mortality. It has since penetrated into many bodies of water (Musselius, 1967), and has been introduced into Sri Lanka (Fernando and Furtado, 1962) and into Romania and other countries of Eastern Europe (Radulescu and Georgescu, 1964).

There are many examples of accidental introduction of pathogenic agents of fish diseases along with fish brought in for selection work. The selection of the cold-resistant strain of carp is based on the crossing of Amur wild carp and the cultured carp (*C. carpio* × *C. carpio hematopterus*). In the course of this work a large number of parasites of Far-Eastern origin came to be introduced into various fish farms.

Back in 1936 the first lot of Amur wild carp was brought in, resulting in the introduction of two monogenetic trematodes which parasitize the gills, and a cestode. *Dactylogyrus extensus* and *Khawiasinensis* cause severe diseases, and at present it is difficult to find a fish farm where these do not occur.

In 1949 some Amur wild carp were introduced into fish farms in the northwestern regions of the USSR and Byelorussia. Some of these fish were infected with the parasitic protozoan, *Ichthyophthirius multifiliis*, which was unknown in the carp ponds of the USSR. Long-distance transportation, which lasted about 20 days, promoted the development of further infection and when transferred to fish farms, the wild carp passed on the infection to the local spawners. In the spring of 1950 a heavy infection caused high mortality among the spawners. At the Velikolusky fish farm 100% of the spawners were lost, and the mortality in the Byelorussian farms was not less than 50%. The disease was introduced with the stocking material into the other farms of Byelorussia and the northwestern regions of the USSR, and by the mid-1950s it had spread to the fish farms of the Ukraine, Central Asia, and Kazakhstan. It was proved experimentally that fish exposed to infection for the first time turned out to be carriers of the largest numbers of *Ichthyophthirius*. The intensity of disease is 20 times less in fish exposed to the infection repeatedly and further exposures helped to develop immunity to the disease (Bauer, 1955, 1958). Toward the end of the 1950s a new lot of the Amur wild carp was introduced into three fish farms in Latvia for selection work. After a year a severe outbreak of dropsy occurred. It was also found that a large nematode, previously unknown in the USSR, was brought in with that lot of fish. Vismanis (1962, 1967) described the nematode as a new species, *Philometra lusiana*. Large females of this nematode, parasitizing under carp scales, cause the formation of ulcers and consequent deterioration of the market quality of fish.

In 1966 a new lot of Amur wild carp was transferred to Ropsha, an experimental fish farm of the State Research Institute of Lake and River Fisheries; every precaution had been taken. In the autumn of 1967, myxosporidians of the genus Sphaerospora were found on the gills of carp in a number of ponds; this parasite had never been found in Ropsha ponds before. It is on record that fish in a number of fish farms were infected by air-bladder disease, introduced there with the carp brood stock brought in for selection purposes. Thus any transfer of fish for the purpose of selection is always accompanied by the danger of introduction of pathogens, even though every precaution is taken. Only introduction of fish at the egg or at the larval stage (during the first few days after hatching) eliminates this danger. This is possible with adequately worked-out methods of artificial spawning. For salmonids and whitefish such methods have been developed since the close of the 20th century and, as a rule, these species are transplanted at the egg stage. Only this may explain the fact that not a single pathogen has been introduced with rainbow trout exported from America to Europe and other continents. At present there are well-developed methods for obtaining fertilized eggs of Acipenseridae and phytophagous Cyprinidae. The eggs of the latter are pelagic. Until recently no effective method is known for taking and incubating adhesive eggs as those of carp. Removal of the adhesive material of carp eggs by any of a number of recently developed methods would enable artificial reproduction of this species, as well as long-distance transportation of the eggs and larvae. In recent years the possibility of obtaining healthy progeny from carp spawners affected by infectious diseases and parasites has been examined. Thus in 1965 (in Byelorussia) and 1967 (in Yazhelbyzy) eggs were obtained from carp spawners affected by air-bladder disease. These eggs were incubated in sterile water, and the larvae were released into ponds where the disease had not been recorded, or to carefully disinfected ponds supplied with water from a noninfected body of water. When mass examination of fingerlings was carried out, not a single individual was found affected by the disease. Artificial reproduction of wild carp has been carried out at the Tsymlansk hatchery-nursery for a number of years and in 1967 about 70 million larvae were obtained. Observations have shown that if larvae are released into carefully disinfected ponds, they will be practically free of parasites when fished out in autumn. Transporting of fish at the egg and larval stages delays the process of obtaining progeny from the transferred fish by several years. But the delay is worthwhile as it excludes the danger of introduction of infectious diseases with fish brought in for selection purposes.

4.2 Deformities as Observed in Different Hatcheries

A large number of specimens having different types of deformities were collected from the nursery and rearing ponds of the study area. The deformities were carefully studied and classified according to the nature of development of deformities involving different parts of the vertebrae, mainly spinal. The maximum numbers of deformities identified were for spinal deformity, followed by stump bodied, and head deformity (Table 4.1 and Figs. 4.1–4.3). The deformed fish were examined under X-radiation and the X-ray photography revealed different

Table 4.1 Type of Deformities Found in Hatcheries of India

SI No	Type of Deformity	No. of Deformities	Average Length (cm)	Average Weight (g)
1	Spinal deformity	15	10.3	23.03
2	Semi operculum	–	–	–
3	Head deformity	3	9.3	18.90
4	Stump body	7	14.8	17.35

Figure 4.1 Sideway curvature indicating scoliosis in Grass carp.

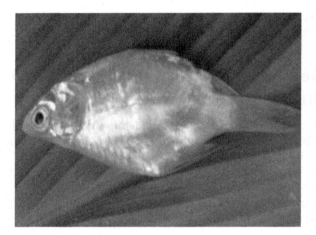

Figure 4.2 Lordosis in Java punti.

Figure 4.3 Lordosis in catla (*Catla catla*).

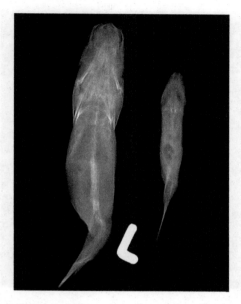

Figure 4.4 X-ray radiograph showing scoliosis (sideway curvature) in Java Punti (*Puntius japonicas*).

categories of vertebral deformities such as kyphosis, lordosis, and scoliosis involving the entire vertebral column (Figs. 4.4–4.7). The spinal column deformities vary with the degree of deformity and with the number of flexions of the vertebral column. X-radiography showed that the entire vertebral column of the affected fish (Fig. 4.4) from cervical regions are irregularly swirled in curves, loops, and incomplete circles resembling a necklace when placed on an uneven ground. Grossly the deformities include exaggerated *kyphosis* (Fig. 4.6), *lordosis*, and *scoliosis* (Figs. 4.4 and 4.7). Individual deformed vertebrae were found to be normal. No erosion, sclerosis, or lytic lesions were observed. It gives an impression of gross spinal dysplasia, possibly due to congenital, hormonal, nutritional, inbreeding causes, etc. The anteroposterior vertebrae are more susceptible to such deformities. X-radiographs of *Catla catla* (Fig. 4.5) and *Puntius japonicas* (Fig. 4.4) revealed similar skeletal deformities, Kyphosis was not observed in *P. japonicas* (Fig. 4.6).

4.2.1 Examination of Deformities

The methodology adopted for examination of deformities is similar to the one followed by Al-Harbi (2001) during a study of cultured common carp *Cyprinus carpio* L. at the Natural Resources and

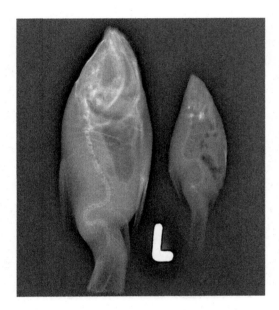

Figure 4.5 X-ray radiograph showing kyphosis in *Catla catla* (curvature at lumbar vertebrae).

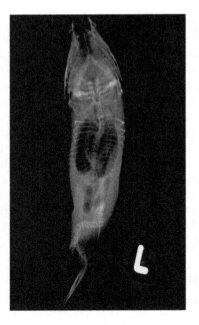

Figure 4.6 X-ray radiographs showing scoliosis.

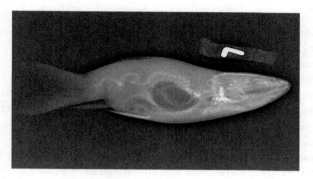

Figure 4.7 X-ray radiograph indicating lordosis in grass carp.

Environmental Research Institute, Saudi Arabia. The specimens collected from the hatcheries located in the study area were counted, measured, and weighed. The fish were then taken to a laboratory to identify the nature and extent of skeletal deformity using a medical X-ray machine. The X-ray photographs were analyzed for examination of the skeletal deformities (Figs. 4.4–4.7).

Skeletal deformities as encountered during study in the three states of India, among the hatchery-raised fishes are mainly three types as mentioned. The incidence of lordosis was reported by Laith et al. (2010) in freshwater mullet (*Liza abu*) collected from Ataturk Lake, Turkey. Three categories of deformities, *lordosis* (ventral deformity, V-shaped), *kyphosis* (dorsal deformity A-shaped), and *scoliosis* (lateral deformity, zigzag-shaped) were collected. This categorization is done depending on the development of deformities at particular region of the vertebral column.

Both the minor and major carp are found to be affected with skeletal deformities, while affected vertebrae are positioned anterior or posterior to the spinal cord. As regards the cultural traits the deformed fishes are inferior to fish having normal vertebrae. To determine the type of deformities, the deformed fish was radioassayed and the study revealed that kyphosis is absent (or may be we did not come across with) in minor carp in the area of study, but fish with lordosis and scoliosis were identified involving both the minor and major carp (Fig. 4.7). Other categories include shortened vertebral bodies and pathological fractures of vertebrae. Among the three types vertebral deformities, *kyphosis* is symptomized as an abnormally rounded upper back (more than 50-degree curvature), starting at the 25th vertebra in the *thoracic region*. Related consequences are reduced rib spaces with subsequent narrowing and collapsing of intervertebral spaces involving vertebrae 18–36 and ultimately leading to the development of asymmetry in concerned intercostal spaces. Abnormal outward curvature of the upper thoracic vertebrae is known as humpback or round back. Kyphosis again is of three main types:

1. Abnormal kyphosis;
2. Postural kyphosis;
3. Congenital kyphosis.

Among the three subtypes, congenital kyphosis differs from the other two as it is caused by an abnormal development of vertebrae during early development prior to birth, and hence is considered genetical in origin.

Lordosis, also known as swayback, in which curvature (which is ventral) is significantly inward and positioned at the lower back involving the lumbar and cervical vertebrae and is known as lordotic. It may occur in any age group and primarily affects the lumbar spine, the affected fish appear swayback. Excessive lordotic curvature is called hyperlordosis, hollow back, and swayback. The number of vertebrae involved is 3–6 and the degree of lordosis is usually evaluated by measuring the angle between the lines passing through the two sides of vertebral column enclosing the curvature. The values of these angles are 133, 84, 105, 92, and 125 (Laith et al., 2010). Unlike other teleost fish, no irregular trajectory of the lateral line was noticed at the position of the moderate curvature (Andrades et al., 1996). Piscine scoliosis is defined as a congenital nutritional deficiency resulting in sideway curvature of the dorsal spine of fish. It is thought that scoliosis in fish is caused by several diverse agents that possibly act on the central nervous system, neuromuscular junctions, or ionic metabolism.

Scoliosis, the abnormal lateral curvature of the spine, originates from the Greek word meaning *crooked*. Developed at an early stage it involves thoracic, lumbar and thoracicolumbar vertebrae transforming it into an S- or C-shaped structure. The severity of the condition is measured in degrees, as viewed from back to front with an X-ray film. In cases up to 25 degrees, doctors monitor the scoliosis for progression, but do not treat it actively. Between 25 and 40 degrees, the standard of care is bracing. Scoliosis is quite a common fish disease, characterized by spine curvature. There is no unanimous opinion about the etiology of this disease, but some scientists consider inbreeding as the causes of various mutations in next generations. Others think that scoliosis is a result of the abnormal development of the ovule caused by feeding female concentrated and unvaried dry food as well as due to various traumas at the larval and fry stages. Other scientists believe that the condition results from the lack of mineral salts or hypoxia. Some have noticed that scoliosis is most commonly diagnosed in the aquaria of selectionists who breed live-bearing species, particularly new species of guppy. A thorough examination of guppy produced by selective breeding showed that spine curvature is observed in all cases of breeding between close relatives. It has also been noticed that the rate of young fish suffering from scoliosis is higher in aquaria with low levels of oxygen. Scoliosis is common in overcrowded rearing, hypoxic water developed out of feeding fish combined dry food, especially in winter, which often results in *bacterium turbidity*. Water hardness in such aquaria ranged from 6 to 12 dh, which rules out the lack of calcium and magnesium salts. Aquarists who feed their fish live food, avoided inbreeding, and maintain a wide variety of plants in the aquarium reported only single cases of scoliosis. These observations and scientific data indicate that inbreeding ischemia may cause scoliosis. Potential causes include varied diet of dry concentrated and combined foods, lack of oxygen, wide range of plants, and overstocking. Traumatic scoliosis occurs very rarely. Malformations of vertebral column, various degrees of reduction of gill cover, deformations of fins, etc., are recorded for many families of carp. Similar abnormalities are often found in the carp of central regions of the Russian Soviet Federal Socialist Republic, North Caucasus and other regions of the former USSR. The deviations of one type vary from 0.1–0.2% to 20–30% or more, the frequency depending on the degree of inbreeding and the environmental conditions under which the fry are reared. In all these cases, we may say that there exists a certain hereditary predisposition to malformations; to be more exact, they are determined both by environment and genotype (Tatarko, 1961, 1966). So far there is no evidential proof of the disease being gene-controlled as well its mode of inheritance. As yet a genetic link has not been identified in carp nor has the mechanism of sex determination yet been determined either, and this may be due mainly to the large number of chromosomes in carp ($2n = 104$). The diploid number of chromosomes of many cyprinids is equal to 50–52, which allows us to consider the common carp as a natural (very ancient) tetraploid (Ohno et al., 1967).

4.2.2 Etiology

Though the etiology of the syndrome (deformity) is not yet understood clearly, a wide spectrum of scientific reasons have been suggested such as: (1) genetic defect, (2) inbreeding depression, (3) temperature fluctuations during early stages of life, (4) rate of water flow in the hatching pool, (5) nutritional deficiency, (6) mineral deficiency, (7) parasitic infection, (8) low water pH, and (9) pollutants and agricultural pesticides.

According to Gjedrem et al. (1991) a high incidence of scoliosis, lordosis, fused vertebrae, curved neural spine, and compressed vertebrae, results from repeated inbreeding—normal practice in the hatcheries of India and some South Asian countries. Nutritional factors, such as vitamin C and D along with minerals like calcium and phosphorus, are also considered for development of spinal malformation, particularly at later stages of development. We know that Ca^{2+} plays an important role in contraction and relaxation of muscle and in fish movement is maneuvered by a v-shaped myotome muscle band being attached to the vertebrae on both sides. Curved vertebrae lead to the, damage and death of a considerable number of myotome muscles, particularly when deformity is in an early developmental, malformation, that leads to a reduced growth rate, due to malformation of locomotory myotome and the associated supporting organ.

Considering the economic importance, due to reduced growth rate and curved phenotypic feature, it fetches a lower price per kg of fish as muscle growth is retarded and in parts muscle even dies (Mehrdad et al., 2011). Some early studies have reported that exposure of malathion in *Heteropneustes fossilis* (Srivastava and Srivatava, 1990), *Brachydanio rerio* (Kumar and Ansari, 1984), and *Cyprinodon variegates* and also exposure of toxaphene in *Pimephalespromelas* (Mehrle Mayer, 1975) and carbaryl in *Orizyias latipes* leads to the development of skeletal and spinal deformities. Recently both copper and zinc pyrithion treatment have been found to develop skeletal deformities in fish. A survey of farmers' needs is required to identify breeding goals and to make sure an improved stock will be well accepted (in view of the traditional preference of locally developed pure breeds in some countries). A national family-based selective breeding program should be started to meet those needs. Dissemination of the improved breed should be done carefully, making sure it does not affect the rare wild resources that may still exist. Biocontainment methods should be applied.

4.3 Interaction Studies

Wild populations, some of which could be 100 years old, dominate the major river systems and other natural habitats. So the efforts to determine the status and genetic structure of feral populations are not so optimistic and are poorly understood. For this reason most of the phylogeographic and population genetic studies are mainly conducted on farmed stock, with occasional involvement of wild stock. Depending on the limited available data it is very difficult to predict on the local adaptation and genetic status of wild populations with varied origin and source. Available data and related information help us to understand the basic biology and distribution of farm-raised common carp. Being a benthic feeder, carp typically make the water turbid. One strong reservation against introduction of common carp to various countries (Australia, United States, Mexico) is that, being a detritovore, the fish turn the water very turbid, while feeding on benthic organisms and putrefied substances. Thus introduced carp create a negative impact on local fish populations and vegetation, which prefer clear water. Apart from incompatibility, another direct effects of the introduction on the aquatic ecology and aquatic species including fish is not yet revealed. Since carp have been introduced to almost all European countries, primarily for recreational purposes, and are already established in open waters, this clearly indicates the potential damage that may have already been done in the aquatic biodiversity. In the backdrop of such a damaging effect, a fresh look should be focused regarding the suitability of carp as an item of introduction, considering all criteria of introduction. Importing countries should impose strict laws regarding such introduction to save native biodiversity.

4.4 Conclusions/Implications: Preservation of Biodiversity

A comprehensive biodiversity survey, aiming at identifying wild populations (native as well as introduced) is urgently needed. It should be carried out in cooperation with local fisheries agencies and/or experts in each country, and should include documentation (including genetic analysis using modern genetic tools to delineate genetic relationships among wild stocks), as well as cryopreservation of semen samples. It is also recommended that areas where wild stocks still exist be declared as sanctuaries to preserve those apparently rare wild resources. These wild populations should be subjected to detailed investigations of life history traits, including reproductive strategies, fecundity, survival, and fitness under variable environmental conditions, e.g., pH and temperature regimes in attempt of identifying local adaptations (e.g., mahaseer).

4.4.1 Unauthorized Introduction by Hatchery Owners and Its Impact on Native Fishery and Aquatic Biodiversity in Bengal, India

In the last 15 years or so the northeastern states have been serving as the gateway for introduction of alien fish species to India from Bangladesh, Thailand, and even South America and Africa. The species chosen for transfer are mainly carnivorous in habit, which includes African catfish (*Clarias garipineus*); Thai catfish (*Clarias macrocephalus*); Pangasid catfish (*Pangasius sutchi*); pacu (*Piaractus brachypomus*); and many others still not known to the scientific community. The introduction of a new species in another environment unit (river, lake, pond, etc.) can be dangerous for local native communities. It is difficult to determine in advance the long range of effects that the introduction of an alien species will have on local communities. Though there exists a prescribed code of practice for alien transfer, it is easier, in the absence of strong rules, to introduce foreign species than to conduct a study on the possible impact of the newcomer in the unit of destination. It is known that introduction of predatory species such as the largemouth bass (*Micropterus salmonoides*) and various salmonids, among others, have caused a number of extinctions worldwide (IUCN, 1985). One most dramatic example of radical extermination caused by alien transfers was the introduction of the Nile perch (*Lates niloticus*) into Lakes Victoria and Kyoga, which caused the extinction of hundreds of endemic species (Barel et al., 1985; Ogutu-ohwayo, 1989). The fish breeders of Bengal have conducted various hybridization programs unscientifically between native and introduced fishes. How many undesirable hybrids are produced through indiscriminate hybridization and the extent of their distribution to diverse geographic territories is not known. The nature of damage already done is also not known in the native fishery due to alien transfer and how many native fish are threatened and/or endangered. Hybridization attempted between nonnative and native species includes virtually all the Indian major carp species with bighead carp (*Aristichthys nobilis*), silver carp (*Hypopthalmicthys molitrix*) and common carp (*Cyprinus carpio* var. Communis). The hybrids produced out of the said outbreeding resulted in the production of some hybrids with some distinctly different morphological characteristics such as almost complete absence of scale or a single row of enlarged scales in some pockets (Fig. 4.9).

4.5 Genetic Concerns With Stock Transfers

The genetic consequences of alien transfer on native fish fauna are of two types. One is direct effects, that results from hybridization and another is indirect effects, which result from factors such as competition, predation, and disease transfer. Stock transfers that lead to hybridization can affect biodiversity in two ways:
1. Through reduction in levels of differentiation between populations;
2. Through reduction in fitness within populations.

4.6 Differences Between Populations

Hybridization increases the heterozygosity or average gene diversity within hybridizing populations and at the same time results in a relative loss of genetic diversity between populations. The prime concern is that locally adapted populations may be replaced by a smaller number of relatively homogeneous ones. This process of consolidation and homogenization may limit the evolutionary potential of the species as a whole. We do not know which alleles or populations may be significant in future so it is important to avoid actions that may reduce the genetic diversity of wild stocks.

4.7 Population Fitness

Population admixtures following hybridization may also affect fitness and may be either positive or negative. This can be viewed along a continuum of the breeding system. At one end assertive breeding, small population size, or both lead to strong inbreeding which ultimately leads to genetic consequences such as inbreeding depression and reduction in fitness. Outbreeding, i.e., hybridization between two different populations or unrelated species, can reduce inbreeding depression. Outbreeding reduces inbreeding depression in two ways:
1. By masking deleterious alleles;
2. By increasing overall fitness through heterosis.

Instead of these positive impacts outbreeding also creates outbreeding depression due to:
1. Loss of local adaptation;
2. Breakdown of coadapted genes at different loci.

Outbreeding depression occurs when there exist differences between the hybridizing populations or species that are genetically based. In the case of hybridization of an alien species with a native one, the hybrids may exhibit reduced fitness to either of the environments as they are not adapted to either of the parental environments. Such a situation resulted in outbreeding depression mainly due to increasing frequency of maladapted genes in the population. Increased fitness also depends on favorable interactions in "coadapted gene complexes" (Dobzhansky, 1955) that evolve to function efficiently as a unit. Outbreeding depression can result from the breakup of these favorable genetic combinations. The effect of stock transfer on the fitness of a population can be gauged by determining the relationship between fitness and degree of outcrossing and the place on the breeding system continuum where a hybrid population will fall. Hybridization resulting from stock transfer can be detrimental as it shifts the hybridizing population toward the right of the breeding continuum, potentially putting the hybrids lower on the fitness scale.

4.8 Effect of Stock Transfer on Fish Biodiversity

Protein electrophoresis and nested gene diversity analysis (Gyllensten, 1985) suggest that the possibility for adverse genetic consequences of stock transfers is greater for freshwater fish (29.4%)

compare to marine fish (1.6%). Experience with two subspecies of large-mouth bass and their hybrids, when they are raised in the native environment of northern species, showed that there is maximum probability for the introduced species to contribute to the extensive "hybrid zone." Hybridization, even at species and subspecies levels, may result in the virtual disappearance of pure native forms, but the introduced species and the hybrids registered a reduced growth rate compared to natives.

4.9 Possible Impact of the Hatchery-Raised Fish in Wild

Supplementation of hatchery-raised fish seed into the natural habitat to increase individual population size, through ranching, is a common practice nowadays. Due to lack of scientific investigation, for most ranching activities it is usually not known whether the transplanted fish replace the native population, hybridize with them, or have no permanent effect on them. According to Hindar et al. (1991) each of the three outcomes is possible, as indicated in Fig. 4.8.

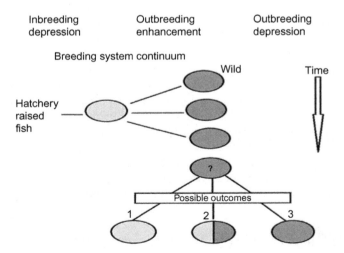

Figure 4.8 Three possible outcomes result from the supplementation of hatchery-raised fish into the wild: (1) total replacement of wild gene pool by hatchery-raised fish; (2) integration (coexistence or hybridization) of native and hatchery gene pools; and (3) persistence of native gene pool with little or no permanent genetic effect of hatchery stock.

Figure 4.9 Hatchery produced hybrid by crossing catla♀ and common (mirror) carp♂, scale developed in a linear row along the middle of the body and in some pockets with an indication of dermal origin.

4.10 Final Impact

Stock transfer can have detrimental effects on aquatic biodiversity by way of reducing between-population variability and within-population fitness. Concerning the negative impact of stock transfer on native fishery, the following steps may be adopted:

1. Stock transfer should be considered only after a full, open, and scientific evaluation of risks involved.
2. Stock transfer, even with positive and/or neutral short-term effects, may have substantial negative long-term effects on the resources.
3. More observations, analysis, thoughts, work plans, and research needs to be devoted to the misuse of defining appropriate units for conservation.

4.11 Extract

The Indian major carp, catla, rohu, and mrigal are the natural inhabitants of the perennial river network of India, Pakistan, and Bangladesh and enjoy a wide distribution. These species are well established in almost all the rivers and reservoirs where they have been transplanted. Substantial research has been carried out to study the biological aspects and propagation of these carp after the initiation of systematic research on inland fisheries at the beginning of the 1950s. Based on this information culture technologies were developed resulting in intensive and extensive composite (multispecies) culture; production levels increased from 0.6 to 10–15 t/ha/year. At the same time three decades of effort to develop improved varieties of carps through cross-breeding (hybridization) did not result in the production of any commercially viable hybrid varieties. Cytogenetic investigations, mainly on karyotypes, have shown that though these major carp have their own distinct morphological features, genetically they appear to be very closely related as indicated by not only their karyotypes but also by their isozyme and DNA profiles. The information available on biochemical genetic studies is not very complete. However, recent efforts in this area and in DNA analyses have resulted in the development of some methodologies for the identification of genetic markers in the case of catla, rohu, and mrigal. These techniques may be effectively used for stock identification and to study genetic variation in the individuals of a particular species or putative landraces. These methods were hitherto not developed and consequently there is no record of previous studies to investigate whether there exist different populations or races within a given species of major carp. With the development of allozyme, RAPD, mtDNA RFLP, and microsatellite techniques, some preliminary investigations were carried out which indicated the existence of genetic variations between the stock of different major seed-producing hatcheries and also between the hatchery and wild populations of catla.

Differences in isozyme markers have been also noticed among rohu of different river systems, namely, the rivers Ganga, Gomati, Yamuna, Sutlej, and Brahmaputra. Of the three species of Indian major carp, rohu grows relatively slower in carp culture systems, while catla and mrigal exhibit comparatively better growth. Because rohu is a highly prized fish, efforts are being initiated to improve the species genetically through selection. There appears to be regional preference, for a particular species in India. While rohu is preferred in most of the states in the Eastern region, catla is popular in the Southern states, especially in the provinces of Karnataka and Andhra Pradesh. Genetic improvement of catla has been taken up in Karnataka province. The aim of current selection projects is to provide correct breeding procedures to the hatchery managers and to develop genetically superior lines that can produce quality seed which can grow to over a kilogram within 1 year under intensive composite farming system. The slowest growing among the major carp, *Labeo calbasu*, is an upcoming species and is gaining popularity. As mentioned earlier, this species has good resistance against parasite infection and can stand rough handling to a better extent than other major carp species. In almost every carp farming nation in Asia, proper use of the genetic potentials of these species has not been sincerely attempted, probably due to the preoccupation to develop suitable culture technologies with the already-existing resources and the need to meet the immediate and urgent protein requirement of the masses in their respective nations. Now many of these nations in

Asia, after achieving the goal to a notable extent in rising the average fish production levels through improved culture technologies, have started realizing that quality improvement is necessary to increase further the quantity. These two aspects cannot be separated. During the coming decades, if aquaculture is to be a viable industry, it must be supported by technologies which ensure sustainable higher productions. Conservation and protection of the genetic resources and exploitation of genetic potentials of the species of fish to their optimum levels appears to be an excellent immediate pathway to achieve the goal.

Most nations, especially in Asia, have already initiated action by taking up conservation of genetic resources and genetic improvement activities of important species. The major carp culturing nations of India, Bangladesh, Pakistan, China, Thailand, Vietnam, etc., have already started working on the job either independently or in a collaborative manner among themselves or with other advanced nations. The International Network for Genetics in Aquaculture (INGA) is coordinating fish genetic research among all the member countries. The above-mentioned countries are among the member countries under the INGA network. India, Philippines, and Vietnam have collaborated with the Institute of Aquaculture Research (AKVAFORSK), Norway, on selective breeding programs. Recently the International Centre for Living Aquatic Resource Management (ICLARM) launched a carp genetic improvement project in which the research on genetic improvement of carp is supported through providing funds and also coordinating the research among the INGA member countries of Bangladesh, China, India, Indonesia, Thailand, and Vietnam. Exchange of carp germplasm among interested countries has also been planned under this program. The coming decade may witness an overall improvement in the aquaculture sector through the systematic and sustainable exploitation of fish genetic resources and the wider use of genetically improved varieties in aquaculture systems.

4.12 Genetic Variance and Breeding

The key component for exploiting the desirable cultural traits of interest in a breeding program is the genetic variance—develop as an outcome of interactions among several components of genetic phenomena, like epistatic effects, additive effects, and dominance effects. Generally, compared to epistatic one additive and dominant genetic variations are more effective as the final outcome is the result of multiple interactions, which are difficult to follow. The relative contribution of either of these effects on the phenotype of interest determines the best approach to use in a breeding program.

While additive impacts are inherited to offspring, dominant effects are expressed in generations. When additive effects are large, the variation within a population will be large also and then fish with a desirable trait can be selected and bred. If the additive variance is small, then selective breeding will not produce any significant effect on offspring. In this case, a recombination of alleles would be the expected approach in the breeding program. More alleles are introduced through an influx of new genetic material, based on trial and error. New strains of fish can be introduced into the breeding program (intraspecific hybridization) or different species bred (interspecific hybridization). A new combination of alleles may produce offspring that harbor the desired combination of phenotypes. For a selection program to be effective, the amount of additive variance should be determined. The measure of heritability (h^2) describes the proportion that additive genetic variance contributes to a phenotype in a population. Heritability describes the percentage of a phenotype that can be inherited in a predicted manner. This value should be reliable since, for a quantitative phenotype, the genotype is not affected by meiosis. These values range from 0 to 1 and are a percentage. A heritability value of 1 suggests that the phenotype observed is 100% explained by the additive genetic variance. A heritability value of 0 suggests that additive genetic effects do not contribute to the phenotype. In general, a heritability value greater than 0.2 ($h^2 = 0.2$) suggests that a trait may be reliably exploited in a selection program. If a trait's heritability is lower than $h^2 = 0.15$, a selection-based program may not be the answer because dominance genetic variance is more important in this case. Just remember that we are dealing with quantitative traits here. These traits may be affected by the environment and the population, and can vary between generations. Direct selection of traits with low inheritance may be difficult; in such situations, genome-based technologies such as marker-assisted selection may be more applicable.

4.13 Genetic Improvement Programs

4.13.1 Selection

Selection is a process in which individuals with desired phenotypes for a particular trait under consideration are identified and used as future brood stock to produce progeny that are also superior for the trait. Selection programs include mass selection and family selection. In mass selection, the performance of all individuals is compared and selection is based on the performance of each, disregarding the parentage. In family selection, the average performance of families is compared and whole families are selected.

Selection programs are reliable only if the genes responsible for the genetic variation are passed to the offspring. Additive genetic variance is transmitted to offspring in a calculated and reliable manner. Heritability (h^2) values must be taken into account. These values are a direct measurement of genetic variance explained by additive genetic effects. The heritability estimates for a trait should be as large as possible to ensure the efficacy of a selection program. Many traits may be correlated, positively or negatively. For instance, fast growth rate is often correlated with a more efficient feed conversion rate in some species (and the opposite can be true in other cases). Therefore, selection for fast-growing fish also selects for fish with a better feed conversion rate. However, if the traits are negatively correlated, the breeder must be careful because selection for one trait may negatively affect the other. For instance, fast growth may be negatively correlated with reproduction capacity so that selecting for fast-growing fish could reduce reproductive capacity. Therefore, in selection programs for certain traits other important traits must be carefully monitored to determine if there is any correlation among the traits.

4.13.2 Hybridization

If heritability is very low, selection methods will not be the best way to increase performance. To improve a trait in this instance, greater genetic variation for new combinations of alleles is needed. New combinations can be created by mating fish with different genetic histories, a process called hybridization. All offspring are called hybrids. The parents can be of the same species, but different strains (intraspecific hybridization). Or, the parents can be of different species (interspecific hybridization). Improving a trait using hybridization is done through trial and error. Some of the hybrids may have superior traits and some may not, but hybridization is the best way to increase performance when the heritability for a trait is very low because hybridization creates new combinations of alleles through dominance genetic variance. This process is independent of heritability, so hybridization programs can be used even when heritability is high. Selection and hybridization can both be used to increase fish performance.

4.13.2.1 Intraspecific Hybridization

Intraspecific hybridization can be used to improve fish performance in one of two ways: It can be used to produce a new strain that can undergo selection for a trait, and it can also be terminal with the hybrids being the end product. In general, hybrids have better fitness because of greater genetic variability. This is known as hybrid vigor or enhancement through outbreeding. To create a new breed, a cross must be made between individuals of two strains with different genetic backgrounds, followed by a selection program to improve performance. A selection program using hybrids usually cannot begin until the second generation of fish (F2) is spawned. This is due largely to the principle of dominance genetic variation; first-generation hybrids (F1) generally do not pass on the hybrid vigor (and superior traits) to all offspring. The hybrids are created using two distinct strains and, therefore, additive genetic effects should be increased. This means that subsequent generations are suitable for use in a selection program. Another approach is to conduct a selection program within a strain and then use hybridization to try to improve performance. However, not all hybrids will show improved performance. Success depends not only on the strains selected but also on the reciprocal hybridization. This means that a male of one strain crossed with a female of a different strain may produce progeny with different traits, and vice versa. The hybrids that perform the best will be determined by experimentation.

4.13.2.2 Interspecific Hybridization

Fish of different species may be crossed to produce more productive progenies. Interspecific hybridization is usually used to exploit hybrid vigor, or the tendency of a crossbred organism to have qualities superior to those of either parent. However, if progenies are fertile, interspecific hybridization has also been used to improve genetics through introgression, a process by which the genes of one species flow into the gene pool of another. This is achieved by backcrossing an interspecific hybrid with one of its parents. The same principles apply as with intraspecific hybridization: Genetic improvement is based on new combinations of alleles. Interspecific hybrids must be able to produce progeny. Once this is established, a breeding program can be attempted. But many between-species hybrids are sterile, do not reproduce as readily as the parents, or produce progenies that are nonviable or abnormal. Spawning does not occur naturally between many species that can be hybridized, so these species must be artificially spawned. The process of producing a superior fish involves experimenting with combinations of strains, species, and reciprocal crosses. An example of successful interspecific hybridization is crossing channel catfish and blue catfish. Each of these species has several different superior traits. Channel catfish is best for commercial production because of its growth rate. Blue catfish has a more uniform body shape, yields more fillet, is easier to seine, and is more resistant to certain diseases. A cross between female channel catfish and male blue catfish produces viable offspring. The hybrids have a faster growth rate and are more disease-resistant than the parental species. With this knowledge, hybridization programs can produce fish that can be selected for further genetic improvements.

4.13.2.3 Polyploidization

Most fish species are diploid; they contain two sets of chromosome pairs ($2n$), one set inherited through the mother and the other set inherited through the father. Polyploids have more than the diploid number of chromosomes. Polyploidy can be induced in fish by using techniques such as temperature variation or pressure applied to the eggs to create triploids ($3n$) or tetraploids ($4n$).

Triploids are created by shocking newly fertilized eggs. The egg does not expel the second polar body when shocked. This creates a fertilized egg with one nuclei from the egg (n), one from the sperm (n), and one from the second polar body (n). During development, the three haploid nuclei will fuse and create a triploid. Triploids also can be created by temperature shocks. Triploids are created to increase fish growth and to control populations by inducing sterility. Triploids should grow larger because the cells are larger (containing more genetic material and larger nuclei). And since triploids are usually sterile, less energy is needed to produce gametes and this energy may be diverted to growth. Triploids are sterile because the normal $2n$ number of chromosomes is disrupted, so that segregation and independent assortment are disrupted. This makes gametogenesis difficult because the chromosomes cannot be divided equally. Many triploids also have abnormal gonads, which makes reproduction difficult. In aquaculture, triploid grass carp are often grown in ponds along with other species to control grasses and weeds. While there is limited use for triploids in large-scale aquaculture, using the process for species with extremely high fecundity, such as oysters, has shown a good level of success. Farmers need large numbers of eggs since many eggs do not survive handling and shocking. Another way to produce triploids is to mate a diploid with a created tetraploid. This method may increase viability, but requires a tetraploid population. Tetraploids have four sets of chromosomes ($4n$). They can be created by shocking a zygote when it is undergoing mitosis. Shock should be applied after the chromosomes have replicated and as the nucleus is about to divide into two. The shock prevents the nucleus and cell from dividing so that it retains four chromosomes. One reason to produce tetraploids is to create triploids, as mentioned above. Triploids can be created more efficiently by mating diploids with tetraploids than they can with shock treatments, because shocking causes significant losses. Many tetraploids can produce viable offspring, so that once a population of tetraploids is created, they could be propagated without creating a new population every time.

4.13.2.4 Production of Gynogens or Androgens

Gynogens are fish that contain chromosomes only from the mother. They are produced by activating oocyte division with irradiated sperm and then restoring diploidy to the developing zygote. Irradiation destroys the DNA in the sperm, but the sperm still can penetrate the egg and induce cell division. After activation of the egg with irradiated sperm, the second polar body normally is extruded, resulting in haploid embryos that eventually die if no additional treatments are given. One way to restore diploidy is to block the extrusion of the second polar body by temperature or pressure shocks. Gynogens so created are called meiogens or meiotic gynogens. Meiotic gynogens are not completely homozygous, even though they contain genetic material only from the mother, because of recombinations between chromosomes in the ovum and in the second polar body. Another way to recover diploidy is to block the first cell cleavage after doubling of the chromosomes. This is done with chemical treatment, temperature shocks, or hydrostatic pressure. Gynogens created this way are called mitogens or mitotic gynogens. Mitotic gynogens also contain genetic material only from the mother and are 100% homozygous. Gynogens can be very useful for genetic studies. They are used to reduce genetic variations and to produce all-female populations in the XY sex-determination system. In certain cases, homozygous genetic material can reduce the complexity of a study. For instance, for whole genome sequencing, a completely homozygous DNA template can reduce the complexities caused by DNA sequence variations between the two sets of chromosomes in regular diploid individuals. Androgens contain chromosomes only from the father and are produced by fertilizing irradiated eggs with regular sperm, followed by doubling of the paternal genome. Fertilizing an irradiated egg with normal sperm produces a haploid zygote. The zygote is shocked after replication, during cleavage, to prevent cell division. The two haploid nuclei fuse together to create a diploid zygote with all-male chromosomal material. This produces two identical copies of haploid male chromosomes. Androgens also can be created by using tetraploid males to fertilize irradiated eggs, which produces diploid offspring. Androgens are completely homozygous and are often referred to as doubled haploid. Androgens are useful for many genetic studies, including reduction of genetic variation and production of all-male or all-female populations. For instance, in the XY sex-determination system, YY males can be produced by androgenesis. Mating YY males with XX females produces an all-male (XY) population. In some cases, as in tilapia production, all-male populations are desirable because of their higher growth rate.

4.13.2.5 Genetic Engineering

Genetic engineering is the process by which a gene(s) or a functional part of a gene is transferred into an organism. The gene may come from the same species as the recipient or a different species. A transgenic fish is produced upon successful gene transfer. The desirable gene is then propagated in the offspring. A number of processes are involved in genetic engineering. First, a gene of interest is cloned and inserted into a vector, such as a bacterial plasmid DNA. The plasmid is then isolated from the bacteria in large quantities. The gene of interest, or the DNA inserted into the plasmid, is removed from the vector and injected into the fish zygote, where it is expected that the new gene will become part of the host fish DNA. Once the transferred DNA is incorporated into the germ cells, it is inherited as a part of the genome; fish so produced are transgenic fish. Genetic engineering has the advantage of breaking the species barrier. It avoids epistasis by transferring specific gene(s). Genetic control elements also can be manipulated to allow a gene to be controlled by a different promoter, as desired. Such promoters can be constitutive or inducible. For instance, a growth hormone gene can be placed under the control of a constitutive promoter leading to constitutive production of elevated growth hormone, which in turn induces fast growth. An inducible promoter can be used to detect specific contaminants for environmental monitoring. For instance, P450 oxidase promoters are inducible upon exposure to certain contaminants. Transgenic fish with these promoters, along with a marker gene, would express the marker gene when the transgenic fish is exposed to the contaminants. In spite of these advantages, genetic engineering is an unconventional approach and the uncertainty of its effect on both food safety and ecological safety has generated much public resistance to its application in aquaculture. To date, transgenic fish have had very limited use and limited economic impact. The well-documented commercial application of transgenic fish is the ornamental glowing zebrafish, the GloFish, created in Singapore and commercially available in the United States. Although the technology for producing transgenic fish is mature, the production and verification process is long and the cost is still high. Major obstacles for using transgenic fish are social resistance and regulations. Therefore, recent studies have focused on assessing transgenic fish in terms of food and ecosystem safety, and on ways to contain transgenic populations.

4.14 Molecular Genetics and Genomics

Molecular genetics is an emerging field in fish breeding programs. It is the study of genetic material (genotypes) to help determine if fish possess certain traits of interest (phenotypes). One such method of genetic testing that will soon become reality for the aquaculturist is DNA marker-assisted selection. When a certain trait of interest is studied, and a genetic marker is found for this trait, a DNA test can determine which fish in the population will be the best to use in a breeding program. Some agricultural programs, such as beef and poultry, have implemented these technologies in their selection and breeding programs already. As our knowledge of the genetics of aquaculture species increases, genetic testing will become a reality for this industry also. The entire DNA composition of an organism is called its genome. Genomics is the study of the entire genome or DNA of a species and how genes interact within the whole organism. Whole genome sequences for many important aquaculture species should be known in the very near future because of advances in sequencing technologies, but the complete DNA sequence of most aquaculture species is presently unknown. Genomic programs may still be useful for any breeding program. Maps of useful traits (their position along the chromosomes) will be valuable in the integration of genomic data with traditional selection programs. Long-term goals of a genomics program would be to identify sets of genes and be able to map multiple production traits to their chromosomes to assist in selection.

4.14.1 Genomic Concepts and Examples of Aquaculture Genomic Research

Genomics is a very active research field. Rather than attempting a thorough review of all the knowledge and progress made, we will discuss some basic concepts of genomics to help readers get to the genomics literature. In aquaculture, major progress has been made in genomics research on many finfish and shellfish species such as Atlantic salmon, rainbow trout, tilapia, carp, striped bass, shrimp, oyster, and scallop. We will conduct further research in catfish as examples for convenience.

4.14.2 Molecular Markers

A first step toward improving aquaculture programs through molecular genetics is to identify molecular markers. DNA sequences vary within a population; that is, alleles at a given locus may be different within populations. Such differences are termed polymorphisms. Identifying polymorphic markers within a species can have commercial importance. For example, a molecular marker can be identified that differentiates a population of slow-growing fish from a population of fast-growing fish based on allele usage. To identify molecular markers, blood or tissue samples may be collected in the field, while DNA isolation and the molecular techniques must be performed in a laboratory. There are many techniques for identifying molecular markers. When no genetic information is available for the aquaculture species of interest, random amplified polymorphic DNA (RAPD) and amplified fragment length polymorphism (AFLP) markers can be used. DNA of the aquatic species is isolated and RAPD or AFLP techniques are used to try to find polymorphisms in the DNA. RAPD and AFLP markers are identified with the help of polymerase chain reactions (PCRs). PCR uses primers (short sequences of synthesized DNA) that bind to DNA and amplify a stretch of DNA between the primer binding sites (Fig. 4.10). In RAPD, short random primers are synthesized and PCR with low annealing temperature is performed on the DNA. The reaction is visualized on a gel and polymorphisms may be identified between DNA samples; they are seen as the presence or absence of an amplified product. The basis of

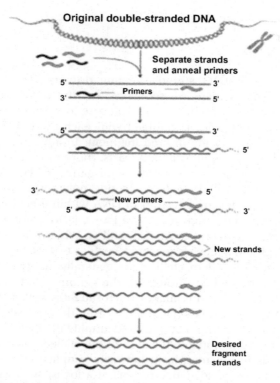

Figure 4.10 Amplification of DNA strand by using primer.

polymorphism in RAPD markers is the differences in DNA sequences between samples. If the DNA sequence at the primer binding site(s) and/or the length of the DNA sequence between the primer sites is different between samples, polymorphisms will likely be observed. The AFLP technique also can identify polymorphic DNA by using PCR. Of the two techniques used to identify molecular markers, AFLP is highly robust and more reliable, but requires more steps than RAPD, as well as some specialized equipment and training. The basis of polymorphism in AFLP is also caused by differences in DNA sequences between samples, observed as the presence or absence of amplified product on a gel. While RAPD and AFLP markers are a quick and economical way to identify polymorphic DNA between samples, they are inherited as dominant markers. Dominant markers, as the name implies, will generate a marker with a single dose of alleles. Thus, with dominant markers, dominant homozygous and heterozygous genotypes are not distinguishable on the basis of the presence or absence of amplified products on a gel. Dominant markers are generally less informative than codominant markers. Communications of dominant markers across laboratories can be difficult as well.

4.15 The PCR

On separation of DNA strands, heating primers are annealed to the targets (complementary) sequence and the DNA sequence between the primers is replicated. The process is repeated and the number of DNA molecules is doubled with each cycle of PCR to produce many copies of the desired fragment. In aquaculture species where some genetic information exists, more molecular markers can be identified. Highly robust and informative markers include microsatellite markers and single-nucleotide polymorphism (SNP) markers. Microsatellites are stretches of DNA within a genome that contain simple sequence repeats. As we know, DNA sequences are composed of four nucleotides: A, T, C, and G. When the combination of nucleotides at a locus is repeated, such as CACACACACACA, a microsatellite exists. Microsatellites are generally highly polymorphic, abundant, distributed throughout the genome, and inherited as codominant markers. This makes microsatellites very useful in developing polymorphic markers. In some cases, the DNA sequences differ at the primer binding site(s) used to amplify the locus by PCR. This would lead to the nonamplification of the allele (so-called null allele). The basis of polymorphism between samples for microsatellites is the number of sequence repeats at a locus, such as $(CA)_8$ versus $(CA)_{10}$. Use of microsatellites as molecular markers requires prior DNA sequence information to identify these repeats, as well as some extra cost and training. SNP markers are codominant markers and using them requires some prior DNA sequence information. SNP markers are also very useful and provide allele-specific information. SNPs are defined as a base change at any given position along the DNA chain (e.g., A–G, or C–T). Theoretically, SNPs should have a total of four alleles (A, T, G, and C at any position); observations suggest, however, that they most often exist as biallelic markers (e.g., the two alleles can be A or G). SNPs can be identified either within an individual (between sister chromosomes in diploid organisms) or between individuals. For example, one allele has the DNA sequence AATAGCTG and another allele has the sequence AATACCTG. In this case, an SNP marker has been identified at that locus. Identified molecular markers can have several uses in genetic analysis. When a polymorphic marker has been identified between populations selected for important traits, individuals can be selected for likely trait performance based upon their genotype (marker-assisted selection). Furthermore, molecular markers can help identify DNA variation useful for inducing new and favorable traits in a selection program where needed. When using an interspecific hybrid system, molecular markers can help identify important genes from each species. Molecular markers can also help identify strain and parentage, and thus confer lineage-specific information (Fig. 4.10).

4.16 Genomic Mapping

Genome mapping techniques include linkage mapping, physical mapping, molecular cytogenetic mapping by fluorescent in situ hybridizations, and radiation hybrid mapping. We will not cover molecular cytogenetic mapping and radiation hybrid mapping here. To use linkage mapping, multiple molecular markers must be identified and a resource family defined. A resource family is a population of individuals (parents and progeny) whose DNA is used for genotyping. Some planning must be done when choosing a resource family for genetic studies. If the resource family is highly informative, a monohybrid cross can be used if parents are true-breeding, or homozygous, for alternate forms of a trait (Fig. 4.4). True-breeding parents produce offspring with a phenotype of interest, say disease resistance. When a species is mated from true-breeding parents (P generation), the offspring are termed the first filial generation, or F1 generation. These F1 progeny are used to create F2 progeny by self-mating, backcrossing, or hybridization. For linkage mapping studies, the F2 generation (and beyond) are chosen as resource families. If the F1 generation is chosen as a resource family, only heterozygous loci of the parents are segregating; the vast majority of homozygous loci of the parents are not segregating. As a result, the F1 resource family may not be fully informative, as all heterozygous siblings will have genotypes with no allele segregation. For instance, if homozygous P fish (AA) and (aa) are mated, all progeny will have the Aa genotype and no allele segregation will be observed. F2 progeny produced by a backcross of F1 fish will produce offspring with the genetic diversity to use as a resource family, as Aa × Aa produces progeny with different genotypes (AA, Aa, and aa alleles are possible). As a practical example, F1 interspecific hybrid catfish (channel catfish × blue catfish) have been created for their superior performance in several commercially important traits. An F2 generation was created by mating F1 generation hybrids with channel catfish and is currently being used as a highly informative resource family.

Linkage maps are created using multiple polymorphic molecular markers within a resource family. Remember, recombination of alleles occurs by crossover of homologous chromosomes. However, recombination frequency is not consistent throughout a genome. Genes on different chromosomes segregate completely independently. When genes are close together on the same chromosome, they are physically linked. These genes are expected to have a lower recombination frequency than genes on the same chromosome but far away from each other. In a resource family, when multiple loci are screened by using multiple molecular markers, linkage maps can be created, usually with the aid of software programs. While the creation of linkage maps is time-consuming and can be complex, the process simply involves the reconstruction of chromosomes (by creating linkage groups) using the recombination differences of molecular markers. When the recombination frequency between markers is low (near 0%), it is expected that the markers are linked and little or no recombination has occurred. Conversely, when recombination frequency between markers is high (approaching 50%), it is expected that the markers are independently assorted during meiosis. The creation of linkage groups is very useful in determining the position and order of markers/alleles along chromosomes. Another useful genomic mapping strategy is physical mapping. Physical maps are created with the help of a DNA library. A library, in terms of genetics, is any collection of DNA fragments that have been inserted into a cloning vector for propagation. For physical mapping, the whole genome of a species can be fragmented and used to construct large-insert DNA libraries. One example of a large-insert DNA library is a bacterial artificial chromosome (BAC) library. A BAC library contains many long pieces (~200,000 nucleotides) of genomic DNA of the species of interest. BAC libraries can be used to create physical maps by a technique called DNA fingerprinting. The genomic DNA contained in a BAC library is isolated and fragmented to create "fingerprints," or highly specific DNA patterns based on the nucleotide composition of the DNA sequence. Overlapping fingerprints are then used to reconstruct the DNA, with the goal of creating a map spanning the entire genome of the species of interest. Physical maps are used in whole genome sequencing projects and are a useful resource for many other genomic projects, including further marker and gene identification, whole genome comparisons to other highly characterized species (map-rich or model species) such as zebrafish and Tetraodon, and in map integration projects. When linkage and physical maps are created, the maps can be integrated (aligned) together, which is particularly useful in analyzing quantitative trait loci (QTL). QTLs are regions of DNA where a correlation has been identified with a trait(s) of interest. QTLs can be added to genomic maps. By integrating the linkage and physical maps, a QTL identified within a linkage map can be located along the physical map. If the marker corresponds to a known gene, the function of this gene may be determined. If the QTL occurs along a region of DNA where no function can yet be assigned, maps are especially useful in locating specific regions along the chromosome for further study. QTL analysis forms the basis for marker-assisted selection, where loci that correspond to candidate genes or traits can be used to assist classical breeding programs. Ultimately, the best genome map can be obtained by sequencing the whole genome for the species of interest. Sequencing a whole genome is costly and requires time, effort, and specialized equipment. Even so, the whole genome sequence of a species can provide a wealth of information. Most whole genome sequencing projects to date are performed by shotgun sequencing and/or minimal tiling path sequencing. Shotgun sequencing is done by sequencing many random, short (~500 nucleotides) segments of DNA and assembling the sequence with the help of computers to reconstruct the chromosomes. With minimal tiling path sequencing, DNA is sequenced in an orderly manner and guides are used to help in the reconstruction process. An example of the minimal tiling path method would be using a BAC library to sequence individual BACs and assembling each BAC one by one. Both strategies are effective, and many projects use a combination of the methods to provide a highly accurate genome sequence. Recent progress in whole genome sequencing includes the use of next-generation (second generation and third generation of sequencers) sequencing technologies. Currently, next-generation sequencing can produce hundreds of thousands to tens of millions of DNA sequences from a sequencing reaction. As more sequencing technologies are developed and they become less costly, whole genome sequencing projects for many aquaculture species are expected.

4.17 Gene Expression Studies

Study of gene expression is considered as one of the primary research areas for further development in fish genetics and biotechnology. The central dogma in molecular biology is the flow of genetic information—DNA → RNA → proteins. Genetic information in coded form on DNA is transcribed into mRNA which translates and converts genetic information to proteins, the end product of protein synthesis hierarchy. The level of expression of a given gene has a direct relation to the amount of mRNA present for that gene. When candidate genes have been identified, gene expression studies can help determine where (what cell or tissue) and how much (quantity of mRNA present) of the gene has been expressed. There are many reasons to perform gene expression studies. An example would be determining gene expression in a control versus a treatment group, such as in healthy fish versus diseased fish. Several genomic tools can help in gene expression studies. The most common technique to identify the expression of individual genes is reverse transcription PCR (RT-PCR). PCR requires DNA templates, so for gene expression studies mRNA must be converted to complementary DNA, or cDNA, using reverse transcriptase, known as RT-PCR. Once mRNA is converted to cDNA, the PCR process works the same way as regular PCR. The idea is that if the starting material contains more of a specific mRNA, more PCR products will be generated using specific PCR primers than when the starting material contains less of the mRNA. This type of RT-PCR is sometimes referred to as semiquantitative PCR because the quantification is not always perfect. Quantification of starting mRNA relies on the PCR to be conducted under identical conditions and stays within the log phase of PCR amplification. A better approach is to use real-time (RT)-PCR. It provides a highly accurate assessment of gene expression levels but is costly and requires specific equipment. Gene expression profiling is used to assess gene expression at the genome scale. When working with a species for which little or no genetic data are available, cDNA libraries are useful in gene discovery projects. Not all genomic DNA corresponds to the gene coding (protein coding) sequence. Therefore, by using a cDNA library for sequencing (instead of genomic DNA), the sequences correspond to gene coding products. Single-run sequencing does not guarantee that the complete mRNA (cDNA) will be sequenced. Usually, only a short fragment of the complete cDNA is sequenced. Single-run sequences of cDNA are called expressed sequence tags (ESTs). Many gene products in an organism can be discovered by generating a large set of ESTs. The frequency of ESTs in a large-scale EST sequencing project is a rough reflection of the gene expression patterns of the organism. However, repeated sequencing of the most abundantly expressed transcripts prevents the rarely expressed genes from being sequenced at all. To create a highly efficient cDNA library for gene discovery projects, multiple sources (cells and tissues) can be used. To circumvent the problem of repeated sequencing of the most highly expressed genes, various types of DNA libraries can be created, including normalized and subtracted libraries. Normalized and subtracted cDNA libraries are often created in the effort to sequence the greatest number of genes. Since cDNAs in a library are sequenced at random, gene products will be sequenced at a frequency relative to their expression levels represented in the library. A normalized library maximizes gene discovery because all genes represented in the library are at a more equal abundance (in theory, at least). Subtracted libraries are useful when comparing two or more expression groups. Subtracted libraries eliminate the cDNA shared by the groups. The remaining cDNA corresponds to the genes that are expressed in one group but not the other(s). The traditional Sanger sequencing method has limitations. The highly efficient next-generation sequencers have the ability to discover the entire mRNA composition of an organism without any subtraction or normalization, which greatly facilitates genome-scale gene expression studies. Once cDNA libraries have been made and ESTs generated, these resources can be used to develop microarrays. In the absence of a whole genome sequence in most of the aquaculture species today, microarrays are often created using all available EST sequences for a species of interest. Common ESTs are combined (clustered) to create a unique set of sequences. These sequences are used to synthesize DNA "features." Features are applied to a medium to create an array of sequences. These features can detect the presence and level of expressed genes in a population of cDNA through hybridization. Microarrays are very powerful tools, limited only by the number and quality of EST resources available with which to design features. Many studies involving microarray technology have been and are currently being performed in aquaculture species, and there can be many applications. One general application would be to determine which genes are differentially expressed in a control group versus a treatment group. This is generally performed using a specific tissue or cell type. An example would be to study the expression profile of healthy fish liver tissue (control) versus liver from a diseased fish (treatment) to attempt to discover genes involved in disease resistance or disease susceptibility. In this example, when cDNA is created from both tissues, the microarray can be used to detect the genes that are upregulated or downregulated between the tissues. These data are useful in determining a global gene expression profile between treatment groups, and also in further research to produce species with superior performance traits.

An understanding of the principles of genetics is useful in any aquaculture program. Genetics programs can help increase productivity in aquaculture systems, as in using hybridization and selection to produce strains with superior performance traits. Genetics research has improved the quality and production of aquatic species throughout the years, but there is a need for further and faster genetic gains. Much of the progress to date has been made using traditional selection. As molecular genetics and genomics tools and technologies are implemented in aquaculture systems, they will complement and extend classical genetic improvement programs.

References

Al-Harbi, A.H., 2001. Skeletal deformities in cultured O. Saleh and T. Borhan, 2009. Identification of common carp *Cyprinus carpio* L. Asian Fisheries Sci. 14 (3), 247–254.

Andrades, J., Becerra, J., Fernandez-Llebrez, P., 1996. Skeletal deformities in larval, juvenile and adult stages of cultured gilthead sea bream (*Sparus aurata* L.). Aquaculture 141, 1–11.

Aquadro, C.F., Bauer Dumont, V., Reed, F.A., 2001. Genome-wide variation in the human and fruitfly: a comparison. Curr. Opin. Genet. Dev. 11, 627–634. doi:10.1016/S0959-437X(00)00245-248.

Arshaniza, N., 1966. Resistance of the Ropsha carp to air bladder disease. Rybovod. i Ribolov. 6, 21. In Russian.

Avetikyan, B.G., 1959. The fate of the alien antigen in the fish organism. Experimental and Clinical Immunology. Leningrad, 270–8. (In Russian).

Barel, C.D.N., Dorit, R., Greenwood, P., et al., 1985. Destruction of fisheries in Africa's lakes. Nature 315, 19–20.

Bauer, O.N., 1955. Ichthyophthirius and its control in pond fish farms. Izv.gosud. nauchno.-issled.Inst.ozer.rech.ryb.Khoz. 36, 184–223. In Russian.

Bauer, O.N., 1958. Parasitic diseases of cultured fishes and methods and their prevention and treatment. In: Dogiel, V.A., Petrushevski, G.K., Polyanski, Yu, I. (Eds.), Parasitology of fishes. Translated by Z. Kabata 1962. Oliver & Boyd Ltd, Edinburgh, pp. 267–300.

Bauer, O.N., Vladimirov, V.L., Tez, V.I., 1965. Fish immunity (Abstract of report). (In Russian).

Bentsen, H.B., Olesen, I., 2002. Designing aquaculture mass selection programs to avoid high inbreeding rates. Aquaculture 204, 349–359.

Dobzhansky, T., 1955. A review of some fundamental concepts and problems of population genetics. Cold Spring Harb. Symp. Quant. Biol. 20, 1–15.

Elinger, N.R., 1964. Selective breeding of trout for resistance to furunculosis. N.Y. Fish Game J. 11 (2), 78–90.

Falconer, D.S., Mackay, T.F.C., 1996. Introduction to Quantitative Genetics, 4th edn Pearson Education Limited, Essex.

Fernando, C.H., Furtado, Y.Y., 1962. Some studies on helminth parasites of freshwater fishes. Proc.I, Regional Symposium Scientific Knowledge of Tropical Parasites held at the University of Singapore, 5–9.

Futuyma, D.J., 1998. Evolutionary Biology. Sinauer Associates Inc, Sunderland, MA.

Gjedrem, T., Gjøen, H.M., Gjerde, B., 1991. Genetic origin of Norwegian farmed Atlantic salmon. Aquaculture 98, 41–50.

Goncharov, G.D., 1963. Fish immunity to infection. 4 Vses.soveshch.bolezn.ryb., Abstracts of reports (In Russian).

Goncharov, G.D., 1967. A study on the mechanism of immunity of fish in infection. "Metabolism and biochemistry of fish" Izdat."Nauka"301–308, In Russian.

Gyllensten, U., 1985. The genetic structure of fish: differences in the intraspecific distribution of biochemical genetic variation between marine, anadromous, and freshwater fishes. J. Fishery Biol. 28, 691–700.

Hazel, L.N., Lush, J.L., 1942. The efficiency of three methods of selection. J. Hered. 33, 393–399.

Hindar, K., Ryman, N., Utter, F., 1991. Genetic effects of aquaculture on natural fish populations. Aquaculture 98, 259–262.

Hogarth, D.M., 1971. Quantitative inheritance studies in sugarcane II. Correlation and predicted response to selection. Anst. J. Agric. Res. 22, 103–109.

IUCN, 1985. 1985 United Nations List of National Parks and Protected Areas. IUCN, Gland, Switzerland, pp. 174.

Karpenko, I., Sventycki, N., 1961. Infectious dropsy control by means of hybridization. Rybovod. i Rybolov. 4, 25–26.

Kirpichnikov, V.S., et al., 1967. Comparative resistance of different breed groups of carp to dropsy. Genetika 7, 57–70. (In Russian, English summary).

Krantz, G.E., Redecliff, J.M., Heist, C.E., 1964. Immune response of trout to *Aeromonas salmonicida*. I. Development of agglutinating antibodies and protective immunity. Progr. Fish Cult. 26 (1), 3–10.

Kumar, K., Ansari, B.A., 1984. Malathion toxicity: skeletal deformities in Zebra fish (Brachydanio verio, Cyprinidae). Pestic. Sci. 15, 107–111.

Laith, A.J., Sadighzadeh, Z., Valinassab, T., 2010. Malformation of the caudal fin in the freshwater mullet, *Liza abu* (Actinopterygii: Mugilidae) collected from Karkhe River, Iran. Anales de Biología 32, 11–14.

Lukyanenko, V.I., 1965. Natural antibodies of fish. Zool. Zh. 44 (2), 300–303. In Russian.

Lukyanenko, V.I., Mieserova, E.K., 1962. Comparative immunological studies of the complement function of fish blood. Dokl. Akad. Nauk SSSR 146 (4), 971–974. In Russian.

Lund, T., Gjedrem, T., Bentsen, H., Eide, D., Larsen, H., Røed, K., 1995. Genetic variation in immune parameters and associations to survival in Atlantic salmon. J. Fish Biol. 46, 748–758. http://dx.doi.org/10.1111/j.1095-8649.1995.tb01598.x.

Mehrdad, Y., Mehdi, R., Mahsa, A., 2011. A radiographical study on skeletal deformities in cultured rainbow trout (*Oncorhynchus mykiss*). Global Veterinaria 7 (6), 601–604.

Mehrle, P.M., Mayer Jr., F.L., 1975. Toxaphene effects on growth and bone composition of fathead minnows, *Pimephales promelas*. J. Fish. Res. Board Can. 32, 593–598.

Musselius, V.A., 1967. Parasites and diseases of phytophagous fish and their control. Izd. Kols, M., 82. In Russian.

Nielsen, R., 2005. Molecular signatures of natural selection. Annu. Rev. Genet. 39, 197–218. http://dx.doi.org/10.1146/annurev.genet.39.073003.112420.

Ogutu-Ohwayo, R., 1989. The purpose, costs and benefits of fish introductions: with specific reference to the great lakes of Africa. In: Utilisation of the Resources and Conservation of the Great Lakes of Africa, 29 Nov-01 Dec 1989, Bujumbura, Burundi, pp. 1–32.

Ohno, S., et al., 1967. Diploid-tetraploid relationship among old-world members of the fish family Cyprinidae. Chromosoma 23 (1) 9 p.

Radulescu, Y., Georgescu, R., 1964. Noi cercetari asupra infestari pestilor cu *Bothriocephalus gowkongensis* Yeh. Bul. Inst. Cerc. Pisc. 1, 78–86. (In Romanian with French and Russian summaries).

Refstie, T., 1982. Preliminary results: difference between rainbowtrout families in resistance against vibrosis and stress. In: Musiwinkles, W.B. (Ed.), Development and Comparative Immunology. Pergamon Press, New York, NY, pp. 205–209. Suppl. 2.

Røed, K.H., Brun, E., Larsen, H.J., Refstie, T., 1990. The genetic influence on serum haemolytic activity in rainbow trout. Aquaculture 85, 109–117.

Schäperclaus, W., 1930. Pseudomonas punctata als Krankheit-serreger bei Fischen. Z. Fisch. 28, 289–370.

Schäperclaus, W., 1953. Bekampfung der Infectionsen Bauchwassersucht des Karpfens durck Zuchtung erblich widerstandfahiger Karpfenstamme. Z. Fisch., I, N.F.H. 5 (6), 321–353.

Schäperclaus, W., 1954. Fischkrankheiten. Akademieverlag, Berlin.708.

Skinner, I.E., 1971. Anguilla recorded from California. Calif. Fish Game 57 (1), 76–79.

Srivastava, A.K., Srivatava, S.K., 1990. Skeletal anomalies in Indian catfish (*Heteropneustes fossilis*) exposed to malathion. J. Environ. Biol. 11, 45–49.

Tatarko, K.I., 1961. The aberrations in structure of gill cover and fins in carp. Vop. Ikhtiol. 1 (3), 412–420.

Tatarko, K.I., 1966. Anomalies of the carp and their causes. Zool. Zh. 45 (12), 1826–1834. In Russian.

Tomasec, I., et al., 1964. Weiterer Beitrag zur Aetilogie der Infectiosen Bauchwassersucht des Karpfens. Jug. Akad. Znanosti i umjetnosti.

Vismanis, K., 1962. Philometrosis of carp in fish farms of the Latvian Soviet Socialist Republic. Izv. Akad. Nauk Latv. SSR 4, 93–96.

Vladimirov, V.L., 1966. Function of antibody formation in fish and its connection with the humoral factors of natural immunity-complement and lysozyme. Simp. po Parasitam i Bolesnyam Ryb i Vodnykh Besposvonochnykh. Tesicy Dokladov: 10–12 (In Russian).

Wolf, L.E., 1954. Development of disease-resistant strains of fish. Trans. Amer. Fish. Soc. 83, 342–349.

5

NEGATIVE ASPECTS OF BREEDING PRACTICE

Induced Fish Breeding. DOI: http://dx.doi.org/10.1016/B978-0-12-801774-6.00005-5

5.1 Use of Impotent Pituitary Gland

Pituitary gland is generally collected by a gland collector from the local fish market during or just before the breeding season, mainly from dead ice-preserved fish. After collection they sell it to a distributor/bulk supplier who preserves it in a refrigerator. The fish breeders purchase gland from these suppliers. In this supply chain no one is aware of the potency of the gland and they are concerned only with making money. The fish breeders select the gland, for extract preparation, by color. Glands with light-brown color and which exhibit spongy property are selected for extract preparation. The active principle of pituitary gland is GTH, which is glycoprotein in nature and extremely sensitive to temperature denaturation. Again the extract often contains microorganisms, especially when prepared from gland collected from dead preserved fish. Due to reduced potency the injected extract fails to cause complete ovulation and spawning. This ultimately results in reduction in population size and inbreeding. It is evident now that a single cycle of full-sib mating resulted in a 10–20% depression in growth rate with a considerable number of abnormalities.

5.2 Restocking of Brooders

Replacement of stock, which is the primary prerequisite of a quality seed production program, is not practiced at all, not only in India but in most Asian countries. On the contrary, fish breeders, particularly medium and marginal fish breeders, use 70–80% brooders for the subsequent breeding program.

5.3 Genetic Appraisal of Improper Fish Breeding Practices

Breeding is the practical aspects of the science of fish genetics and fish endocrinology. By developing the procedure to breed fish in captivity, ignorant fish breeders have otherwise started playing with fish gene pools. It is noted that due to ignorance, genetic aspects are not properly attended to by breeders. This was first elucidated by two seminal research papers published by Allendorf and Phelps (1980) and Ryman and Ståhl (1980). The adverse impacts due to unscientific fish breeding practices are summarized in the following points.

1. Improper dissemination of technology and absence of any follow-up action thereafter;
2. Use only for profit-making proposition without considering any consequences;
3. Impotency of pituitary gland;

4. Use of immature fish as brooders (below 2 years);
5. Mixed spawning;
6. Inbreeding;
7. Genetic drift;
8. Contamination of the wild gene pool, resulting in genetic homogenization.

All the described unscientific and profit-making endeavors initially reduce the population size which, in the subsequent years, invited numerous genetic phenomena, ultimately posing a threat of extinction to affected species.

5.3.1 Inbreeding

Inbreeding, a regular phenomenon in the present-day hatcheries of India and most other Asian countries, occurs through repetitive breeding by maintaining a small founder population. This results in mating among closely related individuals, such as brother–sister and parent-offspring mating. Inbreeding is a cumulative phenomenon and in the course of successive generations it increases homozygosity by 50% and reduces heterozygosity by 50% in F1, 25% in F2, 12.5% in F3, and 6.25% in F4. Continuous inbreeding results in increased homozygosity and produces homozygous stocks of dominant or recessive genes and eliminates heterozygosity from the population. Table 5.1 indicates how

Table 5.1 Inbreeding Reduces the Level of Heterozygosity. It May Be Noted That the Heterozygosity Level Is Reduced to Half After Each Generation of Inbreeding

Generation	Genotype			Homozygosity Level	Heterozygosity Level
	AA	**Aa**	**aa**	**(%)**	**(%)**
Parental	0	100	0	0	100/2
F1	25	50	25	50	50/2
F2	25 + 12.6 = 37.6	25	25 + 12.6 = 37.6	75	25/2
F3	37.5 + 6.25 = 43.75	12.5	37.5 + 6.25 = 43.75	87.5	12.5/2
F4	43.75 + 3.125 = 46.875	6.25	43.75 + 3.125 = 46.875	87.5 + 6.25 = 93.75	6.25/2

a hypothetical population being 100% heterozygous (A a) at parental generation gradually reduced to 1/8th proportion in three generations out of inbreeding. Compared to other vertebrates, fish are more prone to inbreeding in hatchery environment a due to their high fecundity. For example, the species of Indian major carp can produce about 0.25 million eggs per female per breeding season. Fish breeders, to ensure a high rate of fertilization, hatching, and survivality, practice stripping in most cases. Again, it is evident now that multiple breeding and mixed spawning is a common practice in almost all the hatcheries of the Asian continent. The absence of a scientific brood stock management and development program creates a dire scarcity of brood stock during the breeding season. This led fish breeders, particularly the medium-sized, small, and marginal breeders to practice breeding with very small numbers of founder population, resulting in a high rate of inbreeding. The inbreeding rate in some of these hatcheries has been found to be between 2% and 17% (Eknath and Doyle, 1990).

The effects of inbreeding were examined in several species of fish, including common carp, channel catfish, zebra fish, etc., by sib mating (Bondari and Dunham, 1987; Padhi and Mandal, 1994). Now it is well known that inbreeding leads to a reduction in growth, food conversion efficiency, and survival rate and increased production of abnormal offspring. This phenomenon is called inbreeding depression. For example, one generation of inbreeding in common carp reduced the growth rate by 10–20%. Inbreeding though has very little influence on the change in overall gene frequencies, but has major effects on the frequencies of homozygotes, which are increased. If a recessive gene is rare, it will appear in homozygous condition at a higher frequency during inbreeding than random mating. This facilitates increased selection of rare alleles but on the contrary inbreeding is associated with unfavorable biological effects while crossing with unrelated stock results in increased vigor. Inbreeding is not always bad and can be used wisely for the development of aquaculture. The progeny that are produced are homozygous at one or two loci, and are known as *inbred*. Inbreeding can be used to incorporate into offspring the desirable traits of ancestors, with a view to increase the quantity of favorable traits (alleles) in the gene pool of the concerned population. For the incorporation of such traits the procedure adopted is known as *line breeding*. The interbreeding of individuals, when restricted to a particular line of descent, is usually undertaken to perpetuate desirable characters in the subsequent generation.

This indicates that due to inbreeding in each generation, heterozygosity is reduced by 50% and expected to be eliminated from the inbred line with subsequent production of two homozygous pure lines. This is because heterozygosity possesses several heterozygous allelic pairs and offers an opportunity for inbreeding to operate on all gene loci, resulting in the production of totally pure or homozygous offspring.

Genetic theory predicts that inbreeding unmasks deleterious recessive alleles, and the effects as observed are reduced growth rates, lower fecundity, and high infant mortality, which may ultimately lead to population extinction (Caro and Laurenson, 1994). In a large population with random mating these detrimental alleles exist in heterozygous state and therefore expression of deleterious alleles is partially masked by the dominant one. In small populations however, mating among relatives becomes common and the proportion of individuals that are homologous at many loci increases, which results in inbreeding depressions. The theory predicts inbreeding depression as the real cause of concern for conservation of many species that are already on the path of decline, being reduced to small isolated populations. Researchers have questioned the importance of inbreeding for the persistence of wild populations since most inbreeding evidence comes from domesticated/captive populations. Still opinions are there that most inbreeding depression can be purged by some ways and means to avoid inbreeding (Keller and Waller, 2002). It has been noted by the scientists that the importance of inbreeding depression in the wild is not the prime cause of population decline (Caro and Laurenson, 1994), instead environmental stress such as food and water shortages can reveal effects not visualized under less demanding conditions. Inbreeding depression and its interaction with environmental stress were demonstrated in several experiments (Jimenez et al., 1994). Wild species, when subjected to full-sib mating in captivity, exhibit greater loss in population size and lower survivality when they were returned to their normal habitat compared to the progeny of outbred mating. When an equivalent number of animals is kept under laboratory conditions, inbreeding depression is less. This suggests that the prime effect of inbreeding depression lies in its tendency to aggravate the consequences of environmental downturns. However, several other studies have revealed connections of inbreeding with population decline, which led us to conclude that extinction risk is the direct outcome of inbreeding (Saccheri et al., 1998), while environmental/ecological factors may facilitate the process at the second stage. The above discussion indicates that inbreeding depression is common even in the wild and may create a short-term extinction threat, especially if populations are subjected to environmental stress or rapid population decline. The severity of inbreeding and genetic variability loss can be reduced by immigration and/or ranching of a few individuals into a population—known as the *rescue effect* but in the absence of any such migrants, inbreeding depression may act as a driving force to put the populations to an extinction vortex.

5.3.2 Practical Application of Inbreeding

As inbreeding increases, homozygosity of deleterious recessive genes increases in a cumulative fashion with subsequent production of defective phenotype so, like all other sectors, inbreeding should be avoided. It may be mentioned here that, in human society, marriages among close relatives have been banned. Inbreeding can be used to incorporate into offspring the desirable traits of ancestors, with a view to increase the quantity of favorable traits (alleles) in the gene pool of the population concerned. For incorporation of such traits the procedure adopted is known as line breeding (Fig. 5.1). The inter-breeding of individuals within a particular line of descent is usually to perpetuate desirable characters in the subsequent generation.

5.3.3 Inbreeding Depression

Inbreeding, if it continues for generations, leads to the loss of vigor, i.e., reduced growth rate and conversion efficiency, less spawn and fry survival, disease susceptibility, and various deformities. It is measured as the decline in the performance of quantitative traits from the population. In rainbow trout, it has been experimentally proved that one-generation sib-mating resulted in increased fry deformities (37.6%), and decreased food conversion efficiency (15.6%) and fry survival (19%). Again, in common carp, one-generation sib-mating results in a 10–20% reduction in growth rate.

For a given locus, some alleles confer more fitness on an individual than other alleles. Within the "other" class of alleles are rare deleterious recessive alleles, which when appearing as a homozygous genotype in an individual because of mating between relatives, greatly reduces the fitness of the individuals carrying them. Deleterious alleles develop constantly through mutation, so they are always present in a

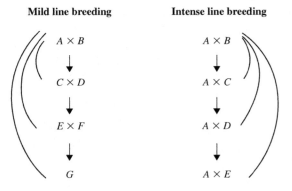

Mild line breeding   Intense line breeding

$A \times B$ $A \times B$

$C \times D$ $A \times C$

$E \times F$ $A \times D$

G $A \times E$

Figure 5.1 Two forms of line breeding.

population at low frequencies. Suppose two alleles, "A" and "a," where "A" is normal allele and "a" is a deleterious allele. Now, if "p" is frequency of "A" and "q" is the frequency of "a," then the homozygous genotype "aa" of the deleterious allele will be rare in a large population, because in random mating the expected frequency of a homozygote will be the square of the allelic frequency p^2, and for a low-frequency allele this is a small value. AA individuals are the most fit of the three possible genotypes (AA, Aa, aa). "A" being dominant, the fitness of individuals possessing genotype AA and Aa will have optimum level of fitness, or they may have some intermediate level of fitness if the effects of the alleles are more additive. Lastly, aa individuals have a recessive deleterious trait that reduces their fitness. In case of two alleles, if one genotype (AA or Aa) produces more offspring than the other in the same environment, then the dominant allele is considered more fit and the phenomenon is called fitness, adaptive value, or selective value.

In a large randomly mating population, the allele "a" occurs at a low frequency and in heterozygous condition, i.e., "A a." In such cases the deleterious allele "a" in the phenotype is masked by dominant "A" allele. But if mating occurs between relatives in which both relatives have a copy of the deleterious allele in the heterozygous state, an Aa × Aa mating, one-fourth of the offspring will carry the deleterious "aa" genotype (Fig. 5.2). This indicates mating between relatives having heterozygous genotype "unmask" the deleterious effects of recessive deleterious alleles that would otherwise remain masked in heterozygous individuals. When deleterious alleles are present in greater numbers and if the offspring are homozygous (aa) due to repetitive inbreeding, then the fitness reduced to the level of 200% with consequent mass mortality of fish.

Now, if the frequencies of recessive homozygotes of an inbred population are compared with a random outbred population, it will be observed that inbreeding changes frequencies of genotype. In a population at equilibrium with genotype and gene frequencies AA = p^2, Aa = $2pq$, and aa = q^2 changes like the following:

$$AA = P^2 + 2pqF : Aa = 2pq + 2pqF; aa = q^2 + pqF$$

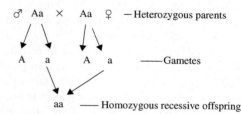

Figure 5.2 Cross of two heterozygous individuals, carrying deleterious "a" allele, produces homozygous "aa" offspring with consequent unmasking of recessive "a."

Here "F" is the inbreeding coefficient or the measurement used for the probability that two genes (in the same locus) in a zygote are identical. It is clear from the discussion that if a disease is controlled by homozygous recessive gene aa with frequency of q^2 in a random outbred population, its frequency will increase by pqF in an inbred population. The ratio of frequencies of recessive homozygotes in inbred and outbred populations will be:

$$= \frac{q^2 + pqFq(q + pF) = q + pF}{q^2 q}$$

The above ratio indicates that if q is large p will be small and F is small also, so the inbreeding increment will be relatively small and increase in recessive homozygotes will be negligible. But when q is small and p is large then pF increases remarkably, the increase of recessive homozygotes, even F, is fairly small. The various ways to alleviate inbreeding are (1) increment of population size, (2) by avoiding mating of close relatives, (3) implementation of outbreeding, i.e., crossing of two genetically distinct populations because outbreeding reduces inbreeding depression by masking the deleterious recessive allele with a dominant normal allele to be supplied by another, genetically different population, or (4) by increasing overall fitness through heterosis or hybrid vigor. It should be kept in mind that too little or too much genetic variability among parents can break down the fitness of their descendants. These pernicious effects are known as "inbreeding depression" when they result from reproduction among related individuals, and "outbreeding depression" when parents are too genetically distant.

5.3.4 Outbreeding Enhancement or Heterosis

Heterosis refers to the phenomenon that progeny of diverse varieties of a species or crosses between species exhibit greater biomass, speed of development, and fertility than both parents. Various models have been posited to explain heterosis, including dominance, overdominance, and pseudo-overdominance. Heterosis, otherwise known as outbreeding enhancement, is the opposite of inbreeding depression and is sometimes referred to as hybrid vigor. Hybrid vigor is the opposite of inbreeding depression and is described as "the masking of recessives through crossing of unrelated genotypes." When two different populations of a species, carrying different recessive deleterious alleles, are crossed, then the probability of the hybrid being homozygous for the same deleterious alleles is not feasible. On the contrary, the offspring will be more fit than either parent due to masking of recessive deleterious alleles by a dominant one. In the subsequent generations when the offspring are allowed to mate randomly, the deleterious alleles will segregate out, following Mendelian inheritance, and will produce individuals with homozygous deleterious alleles with reduced fitness. But the mean level of fitness in the population will still be higher than that of either parental population, because the frequency of each deleterious allele would be reduced by mixing.

In a country, regional populations generally possess dissimilar recessive deleterious alleles in their genetic composition. Mating from two populations may create offspring that are heterozygous for them. The hybrids benefit as any deleterious alleles are masked by heterosis in a recessive state. In successive generations a higher measure of robustness is evident in the nucleus population, but Mendelian recombination will again turn out homozygous deleterious alleles in certain individuals. This principle will hold true in most situations when limited to alleles on a single locus.

If, in a scientific breeding program, inbreeding depression and outbreeding enhancement are the primary genetic mechanisms to be considered and mating between individuals are to be controlled, then the best strategy would obviously be to conduct breeding between individuals from different populations. However, things are not so simple as it appears. Apparently it appears that genes or alleles of only a single locus produce effects toward fitness, but typically alleles lodged at different loci interact, resulting in the creation of gene complexes in a population, and due to uniform and harmonious interactions of these alleles with one another a high level of fitness will evolve. Different isolated populations may evolve different complexes of genes that interact well within a particular population, but cross-population mating resulted in reduced fitness. This reduction in fitness in the offspring is called outbreeding depression.

5.3.5 Genetic Cause of Heterosis

There are two competing hypotheses, developed by Crow (1948), toward explanation of the cause as follows.

5.3.6 Dominance Hypothesis

The dominance hypothesis attributes the superiority of hybrids to the suppression of undesirable recessive alleles from one parent by dominant alleles from the other. It attributes the poor performance of inbred strains to the loss of genetic diversity, with the strains becoming purely homozygous at many loci. The dominance hypothesis was first expressed in 1908 by the geneticist Charles Davenport.

5.3.7 Overdominance Hypothesis

Certain combinations of alleles that can be obtained by crossing two inbred strains are advantageous in the heterozygote. The overdominance hypothesis attributes the heterozygote advantage to the survival of many alleles that are recessive and harmful in homozygotes. It attributes the poor performance of inbred strains to a high percentage of these harmful recessives. The overdominance hypothesis was developed independently by East (1908) and George Shull (1909). Dominance and overdominance have different consequences for the gene expression profile of the individuals. If overdominance is the main cause for the fitness advantages of heterosis, then there should be an overexpression of certain genes in the heterozygous offspring compared to the homozygous parents. On the other hand, if dominance is the cause, fewer genes should be underexpressed in the heterozygous offspring compared to the parents. Furthermore, for any given gene, the expression should be comparable to the one observed in the fitter of the two parents.

5.3.8 Genetic and Epigenetic Basis of Heterosis

The genetic *dominance hypothesis* attributes the superiority of hybrids to the masking of expression of undesirable (deleterious) recessive alleles from one parent by dominant (usually wildtype) alleles from the other. It attributes the poor performance of inbred strains to the expression of homozygous deleterious recessive alleles. The genetic *overdominance hypothesis* states that some combinations of alleles (which can be obtained by crossing two inbred strains) are especially advantageous when paired in a heterozygous individual. This hypothesis is commonly invoked to explain the persistence of some alleles (most famously the sickle cell trait allele) that are harmful in homozygotes. In normal circumstances, such harmful alleles would be removed from a population through the process of natural selection. Like the dominance hypothesis, it attributes the poor performance of inbred strains to expression of such harmful recessive alleles. In any case, outcross mating provides the benefit of masking deleterious recessive alleles in progeny. This benefit has been proposed to be a major factor in the maintenance of sexual reproduction among eukaryotes, as summarized in the article Evolution of sexual reproduction (Letunic and Bork, 2011). The evolution of sexual reproduction describes how sexually reproducing animals, plants, fungi, and protists evolved from a common ancestor that was a single cell edeukaryotic species (Letunic and Bork, 2011). There are a few species which have secondarily lost the ability to reproduce sexually, such as Bdelloidea and some parthenocarpic plants. The evolution of sex contains two related, yet distinct, themes: its origin and its maintenance.

An epigenetic contribution to heterosis has been established in plants, and it has also been reported in animals. Micro-RNAs (miRNAs), discovered in 1993, are a class of noncoding small RNAs which repress the translation of messenger RNAs (mRNAs) or cause degradation of mRNAs (Togashi and Cox, 2011). In hybrid plants, most miRNAs have nonadditive expression (it might be higher or lower than the levels in the parents). This suggests that the small RNAs are involved in the growth, vigor, and adaptation of hybrids (phys.org, 2015).

It was also shown that hybrid vigor in an allopolyploid hybrid of two *Arabidopsis* species was due to epigenetic control in the upstream regions of two genes, which caused major downstream alteration in chlorophyll and starch accumulation (Letunic and Bork, 2011). The mechanism involves acetylation and/or methylation of specific amino acids in histone H3, a protein closely associated with DNA, which can either activate or repress associated genes.

5.3.9 Outbreeding Depression

Crossing populations from different sources may increase reproductive fitness by increasing heterozygosity and thus prevent the expression of recessive alleles (hybrid vigor), or may decrease fitness because of various genetic incompatibilities between the genes from different populations. If outbred offspring have lower fitness than nonoutbred offspring, it is called *outbreeding depression*. Possible causes of outbreeding depression are:

1. *Ecological mechanism*: This mechanism operates through the development of different adaptations in response to different local environments. When these differently adapted individual from different populations are crossed, it results in the development of progeny not suited to either environment and is usually expressed in *F*1.

2. *Genetic mechanism*: It is established now that different populations develop *coadapted gene complexes* in response to the local environment. Coadapted gene complexes are sets of genes, evolved together to produce fittest phenotypes. Outcrossing individuals from different populations having different coadapted gene complexes could result in the disruption of coadapted genes and subsequently reduce fitness. This can be detected in *F*2.

This phenomenon may occur in two ways. One way is by the *"swamping"* of locally adapted genes in a wild population by straying from, e.g., a hatchery population. In this case, adaptive gene complexes in wild populations are simply being displaced by the immigrant genes that are adapted at the hatchery environment or to some other locality. For example, selection in one population might produce a large body size, whereas in another population small body size might be more advantageous. Gene flow and subsequent interactions between these populations may lead to the development of individuals with intermediate body sizes, which may not be adaptive in the habitat of either population. A second way that facilitates the development of outbreeding depression is through the breakdown of biochemical or physiological compatibilities between genes in the different populations. Within local, isolated populations, alleles are selected for their positive, overall effects on the local genetic background. Due to nonadditive gene action, the same genes may have rather different average effects in different genetic backgrounds, hence leading to the development of potential evolution of locally coadapted gene complexes. Offspring between parents from two different populations may have phenotypes that are not good for any environment. It is important to keep in mind that these two mechanisms of outbreeding depression can be operating at the same time. However, determining which mechanism is more important in a particular population is very difficult.

Recent studies have revised outbreeding depression in a variety of plants, and animals. Tony Goldberg's experiments on largemouth bass (*Micropterus salmoides*), a freshwater fish native to North America, is an example of this phenomenon. Since the 1990s, salmonids were found to be infected by *Rhinovirus*. Goldberg et al. sampled healthy individuals from two freshwater bodies, i.e., Mississippi River and the Great Lakes, and created two distinct genetic lines by conducting crossbreeding among the individual lines in separate experimental ponds. After that Rhinovirus was inoculated into two groups of parents belonging to two different lines (generation P) and also in the first- (F1) and second- (F2) generation hybrids of both lines. After 3 weeks, survivality declined to nearly 30% in *F*2 but was more than 80% in *F*1 and the parental generation. This result clearly indicates that crossing of different genetic lines not only increases the probability of self-reproducing population to import diseases, but also weakens its descendants' resistance to future epidemics at a significant level from *F*2 onward. This investigation indicates that translocation of foreign individuals into a self-reproducing population not only imports diseases, but also weaken its descendants' resistance to future epidemics.

5.3.10 Interaction Between Three Mechanisms Needs Modification in Own Format

It is well established now that outbreeding depression as well as enhancement occur simultaneously in a population receiving an alien gene. As discussed above, mating of two genetically different individuals leads to the development of outbreeding depression out of the breakdown and mismatch of the differently adapted gene complexes. But at the same time outbreeding enhancement is expected to occur due to masking of deleterious recessive alleles. It is known now that which phenomenon will overweigh the other is dependent on the rate of outbreeding, when the rate is low then it will lead to the development enhancement. Conversely it will facilitate depression if the outbreeding is random. Currently scientists are not sure at what genetic distance the detrimental effects of outbreeding depression overpower the beneficial effects of outbreeding enhancement. It is dependent on divergence time and subsequent breakdown of coadapted gene complexes. Again the degree of enhancement must be evaluated experimentally.

5.3.11 Mutational Meltdown

Mutational meltdown is the successive accumulation of deleterious mutations in a small population, resulting in loss of fitness and reduction in population size, which may lead to further accumulation of deleterious mutations due to fixation by genetic drift. In stable environments, mutations with phenotypic effects are usually deleterious since populations tend to be well adapted to their environment (Gaggiotti, 2003). Random mutations are likely to disrupt such environmental adaptations. Selection is efficient in eliminating detrimental mutations (with large effects on fitness) in when N_e is large or moderate. Mild deleterious mutations with selection coefficient $s \leq 1/2\,N_e$ behave as neutral mutations and are therefore difficult to remove (Wright, 1931). When N_e drops to a new value, very small deleterious mutations begin to accumulate after approximately $4/N_e$ generations and can rapidly drive populations to extinction when $N_e < 100$–1000 (Kondrashov, 1995). In small populations, selection is hampered and this increases the effective operation of *genetic drift* by increasing the incidence of fixation of some of the deleterious alleles from mutations. This led to reduced fitness and eventually to extinctions (Muller, 1964). Previously, this was thought to be a problem only in obligatory asexual species as there is no recombination (offspring will have parental mutations as well as newly developed mutations; Muller, 1964), but sexually reproducing species are also at risk of extinction due to mutation accumulation (Lande, 1994). If this process repeats, mutations will accumulate and there will be further declines in fitness and population size forming a positive feedback mechanism called mutational meltdown (Lynch and Gabriel, 1990). Recombination in sexual species can slow down mutational meltdown to some extent, but they are not entirely immune to this phenomenon (Gaggiotti, 2003). Empirical evidence for mutational meltdown is scarce for wild populations, and this threat might have been overestimated as an artifact of how the mutation effects on mean fitness have been modeled (Poon and Otto, 2000). In some experiments with yeast (*Saccharomyces cerevisiae*), Zeyl et al. (2001) used 12 replicates of two isogenic strains of yeast with genome-wide mutation rates that differed by two orders of magnitude to demonstrate mutational meltdown. They used an effective population size of about 250 and after more than 100 daily bottlenecks, the yeast with higher mutation rates declined in size and had two extinctions while the wildtype remained constant. These results support the mutational meltdown model (Zeyl et al., 2001), but it has been criticized because of controversies in measures of per genome mutation rates and mean fitness cost per mutation. These measures are thought to be small (Garcia-Dorado et al., 1999), which makes mutational meltdown less likely or less important for most species. Meltdown models ignore the effect of beneficial and backward mutations. Consideration of these mutations might imply that only very small populations would face the risk of extinction due to genetic stochasticity (Poon and Otto, 2000). Also, new mutations may be compensatory or suppressive, which might restore fitness losses incurred by previous mutations without requiring true reversals (Kimura, 1990). Thus currently it is impossible to give clear evaluation of the importance of the meltdown process. In a positive feedback cycle the accumulation of mutational load causes a decrease in population size with reduced fitness. Computer models also suggest that such a mutational meltdown can lead to population extinction and subsequent accumulation of deleterious mutations due to fixation by genetic drift.

A population experiencing mutational meltdown is trapped in a downward spiral and will lead to extinction if the phenomenon lasts for some time. Usually, the deleterious mutations would simply be selected away, but during mutational meltdown, the number of individuals thus suffering an early death is too large relative to the overall population size so that mortality exceeds the birth rate. The accumulation of mutations in small populations can be divided into three phases. In the second phase a population starts in mutation/selection equilibrium, mutations are fixed at a constant rate through time, and the population size is constant because the fecundity exceeds mortality. However, after a sufficient number of mutations have been fixed in the population, the birth rate is slightly less than the death rate, and the population size begins to decrease. The smaller population size allows for a more rapid fixation of deleterious mutations, and a more rapid decline of population size, etc.

This mechanism creates problems in wild populations, especially to endangered species. If we consider the effective sizes of the affected populations are few, then we know from empirical results from several organisms that deleterious mutations, as mild as they may be in their individual effects, appear at a fairly high rate, i.e., about one deleterious mutation appears per individual per generation, indicating, on average, one deleterious mutation in each fish that was not present in either parent.

As we know, the average reduction in fitness when one of these mutations is made homozygous is only about 2%. Earlier speakers noted that the amount of random genetic drift is inversely proportional to population size, $1/2N_e$. If $1/2N_e$ is larger than the selection coefficient, the efficiency of selection against new mutations is less than the force of random drift for that population size. The result is that the "noise" of random drift will overwhelm natural selection and the new deleterious alleles will accumulate in the populations as though they were neutral alleles, even though they have deleterious effects on the individuals that carry them. Thus, if the selection coefficient is 2%, the effect will be important in populations with effective sizes of 50, or with adult census sizes of a few hundred fish. A rule of thumb is that, in small populations, new, mildly deleterious mutations will accumulate in the population at a rate that is half the mutation rate at the genomic level. Even in the absence of inbreeding depression and outbreeding depression, this accumulation of deleterious mutations will lead to a reduction in fitness of about 1% each generation. Since the effective sizes of many endangered populations of salmon are on the order of 50 or smaller, this is a major potential source of long-term genetic deterioration.

5.3.12 Local Extinction in the Presence of Migration

Inbreeding depression is normally reduced by immigrants that are heterozygous for deleterious recessive mutations (Whitlock et al., 2000), and by heterosis mean fitness of populations may be enhanced. However, outcrossing can reduce mean population fitness if hybridization disrupts coadapted gene complexes or favorable epistatic interactions (outbreeding depression). Few studies have demonstrated outbreeding depression as it requires tracking beyond $F1$ generation. A study of song sparrows (Marr et al., 2002), showed signs of outbreeding depression in the $F2$ generation, and measures of fitness were low in the $F2$ generation of crosses of the tidepool copepod (*Tigriopus californium*) from different populations (Burton, 1990). However, this effect of breaking up coadaptations is only magnified if the genetic distance between the two populations has increased greatly (Edmands, 1999). Thus, the threat of outbreeding might not be very serious in most wild populations since it takes many generations in contrasted environments for genetic distance to be significantly large.

The reduction or increase of fitness in a population after receiving immigrants also depends upon interactions among several genetic and nongenetic factors (degree of epistasis, demography, behavior, environmental, etc.; Tallmon et al., 2004). It might therefore be difficult to predict whether any given immigration event will effect genetic rescue, especially when conservation managers lack understanding of the interactions between the genetic and nongenetic factors. However, Gaggiotti (2003) reviewed studies on plants such as *Lotus scoparius*, *Ipomopsis aggregata*, and *Silene diclinis* and concluded that outbreeding depression may be common in the wild but the potential benefits of outbreeding usually outweigh the threats of outbreeding depression.

5.3.13 Extinction in Metapopulation Context

There has been some theoretical work showing that a metapopulation can be subject to extinctions due to genetic factors. Genetic variation can be lost through population turnover. This would be more pronounced when colonizing propagules are formed by individuals from the same deme than from all extant demes (Maruyama and Kimura, 1980). However, if habitat patches differ in quality (the typical case in source-sink metapopulations), then population turnover may not have large effects (Gaggiotti, 2003). Moreover, sink populations can maintain a large proportion of variation in the presence of migration. The mutational meltdown theory was extended to cover metapopulations (Higgins and Lynch, 2001) using individual-based models with stochastic, demographic, environmental, and genetic factors. They concluded that mutational meltdown may be a significant threat to large metapopulations and would exacerbate the effects of habitat loss or fragmentation on metapopulation viability. However there is little empirical evidence supporting predictions made in these theories (Gaggiotti, 2003).

It is well understood now that genetic threats to populations from inbreeding seem the most likely to exacerbate decline and hasten extinction, especially where the reduction in N_e has been very great and under stressful environmental conditions. Most of the genetic threats take many generations to be detected; caution must therefore be taken when making conclusions from studies because what may be insignificant for now might be a threat in the future. In most systems we do not know the threshold where fitness will be an imminent threat of extinction. Also, selection intensity on particular measures of fitness (or life history traits) can vary over time and space, thus the cumulative effects of selection on multiple traits will interact to produce overall fitness effects. This implies that short-term studies of a few traits might result in misleading conclusions.

The division between demographic, environmental, and genetic is artificial since extinction processes often operate together and their synergy may have a stronger impact, especially for populations of intermediate sizes which were previously thought not to be under extinction risk. For very small populations, extinction risk is more influenced by demographic and ecological stochasticity rather than genetic threats. Neither genetic nor demographic factors per se are responsible for most of the populations decline; they only become important after populations have been driven to very low levels, particularly by human activities. Human disturbances such as poaching, habitat fragmentation, introduction of invasive organisms, and pollution present the greatest challenge to populations in the wild than genetic threats. In populations that are less affected by humans (e.g., Checkerspot butterfly), extinction still result from environmental rather than genetic causes (Ehrlich and Murphy, 1987). Conservation efforts should therefore be distributed proportionally to threats posed by any factor, and as shown by most studies, most genetic threats are currently not priorities for many species.

5.3.14 Empirical Evidence

First of all, virtually every trait that has been examined in a wide variety of species can exhibit inbreeding depression, such as by full-sib mating or by self-fertilization in the case of some plants. Some traits are more susceptible to inbreeding than others, but the fact remains that inbreeding depression occurs in all complex genetic characters. A linear decline in mean fitness with the inbreeding coefficient has been observed in a diverse array of organisms including fruit flies, flour beetles, and many species of mammals (including humans). Because inbreeding depression is linear with the inbreeding coefficient, we can extrapolate to future generations if we know the effects of inbreeding depression in the first few generations of inbreeding.

The second point of particular importance for economically important traits in salmon is that traits most closely related to fitness are the ones that exhibit the most inbreeding depression. This has been observed in numerous species, but the data for fruit flies illustrate this principle very well. For morphological characters, the effects of inbreeding are relatively mild but the greatest changes are observed for primary fitness components, such as reproductive capacity, viability, competitive ability, and so on, and not for characters only remotely related to fitness.

One of the noteworthy observations on inbreeding depression is that most of the studies conducted in the laboratory for documentation of observable results were at the end of the experiment (reviewed in Lynch and Walsh, 1997). When parallel studies were conducted both in the laboratory and in the field under natural conditions, the effects of inbreeding were typically much greater under natural conditions. This finding makes an assertion that negative effects of inbreeding outlined above are conservative.

Evidence for outbreeding depression is much less extensive than evidence for inbreeding depression, but outbreeding depression is nevertheless a general genetic phenomenon. One problem in following outbreeding depression is the number of generations that may occur before outbreeding depression reveals itself. The effects of outbreeding enhancement due to the masking of deleterious alleles and outbreeding depression due to hybrid breakdown may cancel each other in the first generation after crossing individuals from two populations. So the effects of outbreeding depression may not be apparent for a few generations. A few experiments have been done in which reciprocal transplants have been made between plants separated by as little as tens or hundreds of meters. In a study of plants separated by various distances, progeny of crosses between plants separated by 10–30 m showed greater fitness than plants separated by smaller or larger distances (Waser and Price, 1989). Many of these studies show that populations are locally adapted and that outbreeding depression occurs between genetically divergent individuals. Comparable studies in animals are rare, but it is likely that similar results occur in animals. Experiments on marine copepods in intertidal pools show that hybrid individuals between populations some tens of kilometers apart show breakdowns in salinity tolerance, prolonged development, and so on (Burton, 1990). In another study, clones of the micro crustacean *Daphnia* in the same lake show hybrid breakdown (Lynch and Deng, 1994). The overwhelming evidence is that these genetic effects occur in every group of organisms studied, and although not much research has been done on salmon, there is no reason to believe that the genetics of salmon are any different.

5.3.15 Heterosis in the Context of Two Depressions

The term "*heterosis*" is often used in genetics and selective breeding, in which desirable traits are inherited by offspring, while undesirable traits are bred out of the species. In breeding, heterosis refers to the idea that a hybrid has greater genetic strength than organisms of a homogeneous (similar) background. It also refers to the potential to combine the positive traits of the parents into "better" offspring. When offspring are considered to be better, or more fit for survival, than their parents, this is known as *hybrid vigor*. Otherwise hybrid vigor may be termed as the manifestation effect of heterosis. However, crossbred plants or animals are not always better than their parents. It is possible for a hybrid to be less fit for survival, which is called outbreeding depression.

1. *Overdominance hypothesis*

The overdominance hypothesis states that an organism that descends from parents of different genetic backgrounds will have greater resistance to a broader spectrum of potential dangers. On the other hand, an organism that descends from parents of similar genetic backgrounds will have resistance to a narrower spectrum of potential dangers. This concept is related to antibody diversity. Similar to genetic diversity, antibody diversity suggests that offspring from parents of different genetic backgrounds are more fit than those from parents of similar genetic backgrounds. This is because they have a greater ability to produce antibodies that can defend against a wider variety of pathogens, or harmful substances, such as bacteria and viruses. Based on this hypothesis, these organisms are more fit due to greater immunity.

2. *Avoidance of deleterious recessive genes*

According to this hypothesis, an organism that descends from parents of different genetic backgrounds will have fewer harmful recessive genes. An offspring inherits genes from his or her parents. One copy of each gene is inherited from the mother and a second copy is inherited from the father. Each parent can only pass one copy of their genes on to the offspring. Which gene gets passed down is determined purely by chance. When both copies (alleles) of a gene are the same, a person is said to be *homozygous* for that gene. If different alleles of the gene are inherited from each parent, the person is said to be *heterozygous* for that gene. Recessive genetic conditions are caused by a mutation, or defect, in a gene. In order to inherit the condition, a person would have to receive two copies of the defective gene. If a person receives one copy of the defective gene and one normal copy, he or she will not have the condition and is known as a carrier. An organism with genetically dissimilar parents is more likely to have fewer recessive genes than an organism born from closely related parents. Therefore, its decreased number of recessive genes may lead to increased fitness. If overdominance is the main cause of greater fitness, then certain genes should be overly expressed in offspring of genetically dissimilar parents when compared to offspring of closely related parents. If the main cause of fitness is the avoidance of recessive genes, fewer genes should be underexpressed in the offspring of genetically dissimilar parents when compared to their parents.

Heterosis is widely used in agriculture and the breeding of various animals for the food supply as well as in scientific research. Heterosis in crops and livestock is different from genetically modified foods, in which new genetic material is introduced into living organisms.

5.3.16 Application of Heterosis in Fisheries

In the past, heterosis has been manipulated for the purpose of eugenics. Eugenics is the science of improving the genetic composition of a population. It generally refers to humans and has been the source of much controversy and ethical debate. While the philosophical standpoint of eugenics is that it lessens human suffering by preventing the spread of negative genetic traits, it is generally regarded as violating human rights. Historically, proponents of eugenics went so far as to forcefully sterilize individuals thought to have such negative genetic traits. By applying heterosis, scientists can bring about more favorable traits in plants and animals used for human consumption while repressing less favorable ones. The other side of this is that breeding in desirable characteristics and breeding out undesirable ones contribute to a less diverse population. There are also implications for recessive genetic disorders that occur in closely related populations. Tay–Sachs disease, e.g., is a recessive genetic condition that is common among certain groups, such as French Canadian and Ashkenazi Jewish populations. Couples from populations in which there is an increased risk of certain recessive genetic disorders may work with genetic counselors to evaluate the probability of having children with the disorder based on known risk factors. Counselors may also help prospective parents to decide which testing methods are appropriate and how to interpret results, and also help decide whether or not to terminate an affected fetus. It is more than certain that genetic tests cannot guarantee accuracy; there is always a risk of terminating a healthy unborn child. Also, clinical studies have shown that women who have had abortions suffer an increased risk of anxiety, depression, and suicide and are at an increased risk for breast cancer.

5.3.17 Limitations

There are two main hypotheses to explain the "fitness advantage" in heterosis, i.e., the overdominance hypothesis and the avoidance of deleterious recessive genes hypothesis. However, a clear conclusion has not yet been reached.

Inbreeding depression, on the other hand, represents the decrease in fitness as a result of the breeding of organisms of similar genetic backgrounds. With inbreeding, rare genetic diseases become more common among populations. Not only are the chances of genetic defects in offspring increased, but researchers also believe that inbreeding may have long-term effects on health by increasing the risk of late-stage disease, such as obesity, heart disease, and type II diabetes. The opposite of a hybrid is a pure bred and is developed out of breeding two organisms that have similar genetic make up with no outbreeding.

5.4 Genetic Drift

Suppose a fish population consists of a male with gray color bearing bb genotype and five black females with BB genotype. Now assume each female is replaced by her daughter in each successive year. The first year offspring will be gray and Bb in genotype. From the second year onward offspring will be either Bb or bb. Each year the number of Bb will decrease while bb will increase. At some point in the near future all individuals will be bb and the gene for gray will have been lost to the population via genetic drift. Extreme loss in genetic variability is often found in island populations or isolated strains that can create adaptation to a specific environment. Yet it can also reduce the ability to adapt to changes in either environment or preference. This is commonly seen with strains geared toward a limited phenotype.

Genetic variation in guppies is jointly determined by breeder selection, initial genetic variability, naturally occurring genetic drift, and mutation. Genetic drift has a high relative importance in small populations. The negative effects of recessive deleterious alleles may become much more prevalent as the frequency of the recessive allele increases. When the recessive allele replaces the dominant allele (or vice versa) a fixed trait results. The loss of any positive or negative allele is most likely to occur in a small gene pool. The reduction in the number of forms of an allele in the extreme case leads to a monomorphic state where only one form exists. Consequences resulting from a loss of genetic variation include inbreeding depression and/or the inability of a strain to adapt and evolve to changing conditions of its environment. While inbreeding depression has been accepted within domestic strains, scientists are just beginning to substantiate it in selected wild or feral populations. The occurrence of inbreeding depression indicates that the underlying causes are yet to be fully understood by the breeder or are breeder-enhanced by poor mating selection and retention.

Phenotypic characteristics (appearance) are often used to divide individuals into groups. In most carp hatcheries, fish breeders used to maintain limited numbers of brood stock and the entire breeding program is carried out of this small number of brood stocks. This led to the development of a phenomenon known as *genetic drift*. It is a phenomenon which creates random changes in gene frequencies in the founder population, which may not carry some alleles due to *sampling error* (Fig. 5.3). The loss of alleles reduces genetic variance in the hatchery population (Fig. 5.4). Allendorf and Phelps (1980) first addressed this problem of hatchery practices leading to genetic drift in Cutthroat trout *Oncorhinchus clarki* (Wallbum). They showed the loss of alleles due to genetic drift by comparing the allelic frequencies

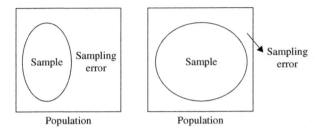

Figure 5.3 Sampling error is less in the case of a large population compared to a small one.

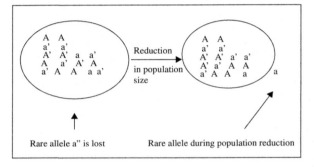

Figure 5.4 Loss of rare alleles by genetic drift is caused by a reduction in population size.

in hatchery-bred fish and their wild relatives. Genetic drift makes a population unfit for selective breeding as occurred in case of *Tilapia nilotica*. It was also found that genetic drift led to the extinction of a certain strain of channel catfish (Tave, 1991).

5.5 Sampling Error

Sampling error is the deviation of the selected sample from the true characteristics, traits, behaviors, qualities, or figures of the entire population.

5.6 Why Does This Error Occur?

Sampling process error occurs because researchers draw different subjects from the same population, but still the subjects have individual differences. Keep in mind that when you take a sample, it is only a subset of the entire population; therefore, there may be a difference between the sample and the population. The most frequent cause of the said error is a biased sampling procedure. Every researcher must seek to establish a sample that is free from bias and is representative of the entire population. In this case, the researcher is able to minimize or eliminate sampling error.

Another possible cause of this error is chance. The process of randomization and probability sampling is done to minimize sampling process error but it is still possible that all the randomized subjects are not representative of the population. The most common result of sampling error is systematic error, wherein the results from the sample differ significantly from the results from the entire population. It follows logic that if the sample is not representative of the entire population, the results from it will most likely differ from the results taken from the entire population.

5.7 Sample Size and Sampling Error

Given two studies the exact same, same sampling methods, same population, then the study with a larger sample size will have less sampling process error compared to the study with the smaller sample size. Keep in mind that as the sample size increases, it approaches the size of the entire population, therefore, it also approaches all the characteristics of the population, thus, decreasing sampling process error.

5.8 Standard Deviation and Sampling Error

Deviations are used to express the variability of the population. More technically, it is the average difference of all the actual scores of the subjects from the mean or average of all the scores. Therefore, if the sample bears high standard deviation then the sampling process error will be high. On the contrary, standard deviation is indirectly proportional to sample size, i.e., as the sample size increases, the standard deviation decreases. Imagine having only 10 subjects, with this very small sample size, the tendency of their results is to vary greatly, thus a high standard deviation. Then, imagine increasing the sample size to 100, the tendency of their scores is to cluster, thus a low standard deviation.

5.9 Random Genetic Drift

Random genetic drift is a stochastic process (by definition). A classic example of random genetic drift is the random nature of transmitting alleles from one generation to the next given that only a fraction of all possible zygotes become mature adults. The easiest case to visualize is the one that involves binomial sampling error. If a pair of diploid sexually reproducing parents (such as humans) have only a small number of offspring then not all of the parent's alleles will be passed on to their progeny due to chance assortment of chromosomes at meiosis. In a large population this will not have much effect in each generation because the random nature of the process will tend to average out. But in a small population the effect could be rapid and significant.

If a population is finite in size (as all populations are) and if a given pair of parents have only a small number of offspring, then even in the absence of all selective forces, the frequency of a gene will not be exactly represented in the next generation because of sampling error. If in a population of 1000 individuals the frequency of "a" is 0.5 in one generation, then it may by chance be 0.493 or 0.505 in the next generation because of the chance of production of a few more or less progeny of each genotype. In the second generation, there is another sampling error based on the new gene frequency, so the frequency of "a" may go from 0.505 to 0.501 or back to 0.498. This process of random fluctuation continues generation after generation, with no force pushing the frequency back to its initial state because the population has no "genetic memory" of its state many generations ago. Each generation is an independent event. The final result of this random change in allele frequency is that the population eventually drifts to $p = 1$ or $p = 0$. After this point, no further change is possible and the population has become homozygous. A different population, isolated from the first, also undergoes this random genetic drift, but it may become homozygous for allele "A," whereas the first population has become homozygous for allele "a." As time goes on, isolated populations diverge from each other, each losing heterozygosity. The variation originally present within populations now appears as variation between populations (Suzuki et al., 1989).

Random genetic drift also refers to accidental random events that influence allele frequency. For example, certain events can cause the frequencies of alleles in a small population to drift randomly from generation to generation. Suppose a population consists of 25 fish, of which 16 are homozygous (AA) for body color, eight are heterozygous (Aa), and one is homozygous recessive (aa). Now suppose three fish that are lost from the population due to death before maturity carry "AA" genotype, then it would certainly alter the relative frequency of two alleles (A and a) for body color in subsequent generations—a phenomenon known as *microevolution* caused by genetic drift.

Another derived phenomenon of genetic drift is *population bottlenecking* which facilitates sharp decline in population size by being influenced by environmental events or human activities (such as genocide). Such events reduce the variation, at the level of the gene pool of a population. After an event, a smaller population (of animals/people), with a correspondingly smaller genetic diversity, remains to pass on genes to future generations. Genetic diversity remains lower, only slowly increasing with time as random mutations occur. In consequence of such population size reductions and the loss of genetic variation, the robustness of the population is reduced; the ability of the population to survive selecting environmental changes, like climate change or a shift in available resources, is reduced. Conversely, depending upon the causes of the bottleneck, the survivors may have been the most fit individuals, hence improving the traits within the gene pool while shrinking it. This genetic drift can change the proportional distribution of an allele by chance and even lead to fixation or loss of alleles. Due to the smaller population size after a bottleneck event, the chance of inbreeding and genetic homogeneity increases and unflavored alleles can accumulate. A slightly different form of a bottleneck can occur if a small group becomes reproductively (e.g., geographically) separated from the main population, such as through a founder event where, e.g., a few members of a species successfully colonize a new isolated island, or from small captive breeding programs such as animals at a zoo. Population bottlenecks play an important role in conservation biology (see minimum viable population size) and in the context of agriculture (biological/pest control). Several examples of bottlenecks have been inferred from genetic data. For example, there is very little genetic variation in the cheetah population. This is consistent with a reduction in the size of the population to only a few individuals—an event that probably occurred several 1000 years ago. An observed example is the northern elephant seal which was hunted almost to extinction. By 1890 there were fewer than 20 animals but the population now numbers more than 30,000. As predicted there is very little genetic variation in the elephant seal population. It is likely that the 20 animals that survived the slaughter were more "lucky" than "fit."

Another example of genetic drift known as the *founder effect*, which depicts that a reduction in genetic variation results when a small subset of a large population is used to establish a new colony. The new population may be very different from the original population, both in terms of its genotypes and phenotypes. In some cases, the founder effect plays a role in the emergence of new species. Initially a small group breaks off from a larger population and forms a new population. This effect is well known in human populations. The founder effect is probably responsible for the virtually complete lack of blood group B in American Indians, whose ancestors arrived in very small numbers across the Bering Strait during the end of the last Ice Age, about 10,000 years ago. More recent examples are seen in religious isolates like the Dunkers and Old Order Amish of North America. These sects were founded by small numbers of migrants from their much larger congregations in central Europe. They have since remained nearly completely closed to immigration from the surrounding American population. As a result, their blood group gene frequencies are quite different from those in the surrounding populations, both in Europe and in North America.

The process of genetic drift should sound familiar. It is, in fact, another way of looking at the inbreeding effect in small populations. Whether regarded as inbreeding or as random sampling of genes, the effect is the same. Populations do not exactly reproduce their genetic constitutions; there is a random component of gene-frequency change. There are many well-studied examples of the founder effect. All of the cattle on Iceland, e.g., are descended from a small group that were brought to the island more than 1000 years ago. The genetic make-up of the Icelandic cattle is now different from that of their cousins in Norway but the differences agree well with those predicted by genetic drift. Similarly, there are many Pacific islands that have been colonized by small numbers of fruit flies (perhaps one female) and the genetics of these populations is consistent with drift models.

These examples, and many more, show that it is wrong to consider natural selection as the only mechanism of evolution and it is also wrong to claim that natural selection is the predominant mechanism. This point is made in many genetics and evolution textbooks. For example, one of the most important and controversial issues in population genetics is concerned with the relative importance of genetic drift and natural selection in determining evolutionary change. The key question at stake is whether the immense genetic variety which is observable in populations of all species is inconsequential to survival and reproduction (i.e., neutral), in which case drift will be the main determinant, or whether most gene substitutions do affect fitness, in which case natural selection is the main driving force. The arguments over this issue have been intense during the past half-century and are little nearer resolution though some would say that the drift case has become progressively stronger. Drift by its very nature cannot be positively demonstrated. To do this it would be necessary to show that selection has definitely not operated, which is impossible. Much indirect evidence has been obtained, however, which purports to favor the drift position. Firstly, and in many ways most persuasively, is the molecular and biochemical evidence (Harrison et al., 1988). The book by Harrison et al. is quite interesting because it goes on for several pages discussing the controversy. The authors point out that it is very difficult to find clear evidence of selection in humans (the sickle cell allele is a notable exception). In fact, it is difficult to find good evidence for selection in most organisms. (All of the arguments are "after the fact," but probably correct!). One of the consequences of fixation by random genetic drift is that beneficial alleles can be lost by drift before they are fixed by natural selection. Take the example of an allele with a 5% selective advantage. The probability of fixation of this allele is only 10% and not 100% as most people imagine. The difference is due to the same stochastic events that govern random genetic drift. In 90% of the cases, the less beneficial allele will reach fixation. Similarly, new detrimental alleles are not always eliminated from the population. In any population, some proportion of loci are fixed at a selectively unfavorable allele because the intensity of selection is insufficient to overcome the random drift to fixation. Very great skepticism should be maintained toward naive theories about evolution that assume that populations always or nearly always reach an optimal constitution under selection. The existence of multiple adaptive peaks and the random fixation of less fit alleles are integral features of the evolutionary process. Natural selection cannot be relied on to produce the best of all possible worlds. The relative importance of drift and selection depends, in part, on estimated population sizes. For any given set of alleles, random genetic drift is much more important in small populations than in large populations. It is important to remember that most species consist of numerous smaller inbreeding populations called "demes." It is these demes that evolve and that make drift very important in the real world.

The effect of population size is not important when you only consider overall change and not the effect on a single allele. This is because larger populations have more mutations and the probability of fixation by random genetic drift is identical to the rate of mutation for most alleles. Thus, large and small populations evolve at the same rate by random genetic drift. Studies of evolution at the molecular level have provided strong support for drift as a major mechanism of evolution. Observed mutations at the level of genes are mostly neutral and not subject to selection. One of the major controversies in evolutionary biology is the neutralist–selectionist debate over the importance of neutral mutations. Since the only way for neutral mutations to become fixed in a population is through random genetic drift, this controversy is actually over the relative importance of drift and natural selection. The founder effect is the reduction in genetic variation that results when a small subset of a large population is used to establish a new colony. The new population may be very different from the original population, both in terms of its genotypes and phenotypes. In some cases, the founder effect plays a significant role in the emergence of new species.

5.10 Ways to Eliminate Sampling Error

There is only one way to eliminate this error by eliminating the concept of sample, and to test the entire population. In most cases this is not possible; consequently, what a researcher can do is to minimize sampling process error and this can be achieved by a proper and unbiased probability sampling and by using a large sample size.

5.11 Simple Measures to Avoid Improper Breeding Practices in the Hatchery

1. The blood stocks should be partially replaced periodically in hatcheries. Exchange of brood stocks between the local hatcheries is a useful way of doing this.
2. Brood stocks of different age groups should be bred together. This helps in reducing the chance of loss of some valuable alleles due to genetic drift.
3. Natural stocks may be inducted periodically to increase the heterozygosity.
4. Cryopreserved spermatozoa may be used, if possible, to maintain heterozygosity in the hatchery population.
5. The pedigree record should be maintained to avoid the mating of close relatives.
6. Crossing of different lines of fishes would increase heterozygosity. Separate lines of fish can be maintained by keeping the record of the families of different strains bred in the hatchery.
7. Indian major carp should be spawned separately in the breeding pool to avoid inadvertent hybridization between these species.

References

Allendorf, F.W., Phelps, S.R., 1980. Loss of genetic variation in hatchery stock of cutthroat trout. Trans. Am. Fish. Soc. 109, 537–543.

Bondari, K., Dunham, R.A., 1987. Effects of inbreeding on economic traits of channel catfish. Theor. Appl. Genet. 74, 1–9.

Burton, R.S., 1990. Hybrid breakdown in physiological response: a mechanistic approach. Evolution 44 (7), 1806–1813.

Caro, T.M., Laurenson, M.K., 1994. Ecological and genetic factors in conservation: a cautionary tale. Science 263 (5146), 485–486.

Crow, J.F., 1948. Alternative hypotheses of hybrid vigor. Genetics 33 (5), 477–487.

East EM (1908). Inbreeding in Corn. Reports of the Connecticut Agricultural Experiments Station for 1907: 419–428.

Edmands, S., 1999. Heterosis and outbreeding depression in interpopulation crosses spanning a wide range of divergence. Evolution 53 (6), 1757–1768.

Ehrlich, P.R., Murphy, D.D., 1987. Conservation lessons from long-term studies of checkerspot butterflies. Conserv. Biol. 1 (2), 122–131.

Eknath, A.E., Doyle, R.W., 1990. Effective population size and rate of inbreeding in aquaculture of Indian major carps. Aquaculture 85, 293–305.

Gaggiotti, O.E., 2003. Genetic threats to population persistence. Annales ZoologiciFennici 40, 155–168.

Garcia-Dorado, A., Lopez-Fanjul, C., Caallero, A., 1999. Properties of spontaneous mutations affecting quantitative traits. Genet. Res. 74, 341–350.

George Shull, 1909. A Pure-Line Method in Corn Breeding. J Hered (1909) os-5 (1), 51–58. http://dx.doi.org/10.1093/jhered/os-5.1.51.

Harrison, G.A., Tanner, J.M., Pilbeam, D.R., Baker, P.T., 1988. Human Biology, 3rd ed. Oxford University Press, New York, NY, pp. 214–215

Higgins, K., Lynch, M., 2001. Metapopulation extinction caused by mutation accumulation. Proc. Natl. Acad. Sci. 98 (5), 2928–2933.

Jimenez, J.A., Hughes, K.A., Alaks, G., Graham, L., Lacy, R.C., 1994. An experimental study of inbreeding depression in a natural habitat. Science 266 (5183), 271–273.

Keller, L.F., Waller, D.M., 2002. Inbreeding effects in wild populations. Trends Ecol. Evol. 17 (5), 230–241.

Kimura, M., 1990. Some models of neutral evolution, compensatory evolution, and the shifting balance process. Theor. Popul. Biol. 37, 150–158.

Kondrashov, A.S., 1995. Contamination of the genome by very slightly deleterious mutations: why have we not died 100 times over. J. Theor. Biol. 175 (4), 583–594.

Lande, R., 1994. Risk of population extinction from fixation of new deleterious mutations. Evolution 48 (5), 1460–1469.

Letunic, I., Bork, P., 2011. Interactive Tree of Life v2: Online annotation and display of phylogenetic trees made easy (PDF). Nucleic Acids Research 39 (Web Server issue): W475-8. http://dx.doi.org/10.1093/nar/gkr201. PMC 3125724. PMID 21470960.

Lynch, M.H.-W., Deng, 1994. Genetic slippage in response to sex. American Naturalist 144, 242–261.

Lynch, M., Gabriel, W., 1990. Mutation load and the survival of small populations. Evolution 44 (7), 1725–1737.

Lynch, M., Walsh, B., 1997. Genetics and Analysis of Quantitative Traits Sinauer Assoc. (December 1997).

Marr, A.B., Keller, L.F., Arcese, P., 2002. Heterosis and outbreeding depression in descendants of natural immigrants to an inbred population of song sparrows (*Melospiza melodia*). Evolution 56 (1), 131–142.

Maruyama, T., Kimura, M., 1980. Genetic variability and effective population size when local extinction and recolonization of subpopulations are frequent. Proc. Natl. Acad. Sci. 77 (11), 6710–6714.

Muller, H.J., 1964. The relation of recombination to mutational advance. Mut. Res. 1 (1), 2–9.

Padhi, B.K., Mandal, R.K., 1994. Improper fish breeding practices and their impact on aquacultureand fish biodiversity. Curr. Sci 68, 965–967.

phys.org (2015). Population benefits of sexual selection explain the existence of males. Report on a study by the University of East Anglia, May 18. http://phys.org/news/2015-05-population-benefits-sexual-males.html.

Poon, A., Otto, S.P., 2000. Compensating for our load of mutations: freezing the meltdown of small populations. Evolution 54 (5), 1467–1479.

Ryman, N., Ståhl, G., 1980. Genetic changes in hatchery stocks of brown trout (*Salmo trutta*). Can. J. Fish. Aquat. Sci. 37, 82–87.

Saccheri, I., Kuussaari, M., Kankare, M., Vikman, P., Fortelius, W., Hanski, I., 1998. Inbreeding and extinction in a butterfly metapopulation. Nature 392, 491–494.

Suzuki, D.T., Griffiths, A.J.F., Miller, J.H., Lewontin, R.C., 1989. An Introduction to Genetic Analysis, 4th ed. W.H. Freeman, New York, NY, p. 704.

Tallmon, D.A., Luikart, G., Waples, R.S., 2004. The alluring simplicity and complex reality of genetic rescue. Trends Ecol. Evol. 19 (9), 489–496.

Tave, D., 1991. Effective breeding number and genetic drift. Aquaculture Magazine, 109–112. Sept/Oct.

Togashi, T., Cox, P. (Eds.), 2011. The Evolution of Anisogamy. Cambridge University Press, Cambridge, p. 22–29.

Waser, N.M., Price, M.V., 1989. Optimal outcrossing in *Ipomopsis aggregata*: seed set and offspring fitness. Evolution 43, 1097–1109.

Whitlock, M.C., Ingvarsson, P.K., Hatfield, T., 2000. Local drift load and the heterosis of interconnected populations. Heredity 84, 452–457.

Wright, S., 1931. Evolution in Mendelian populations. Genetics 16 (2), 97–159. PMid:17246615; PMCid:1201091.

Zeyl, C., Mizesko, M., De Visser, J.A.G.M., 2001. Mutational meltdown in laboratory yeast populations. Evolution 55 (5), 909–917.

6

INFLUENCE OF ECOLOGICAL FACTORS ON MATURATION, SPAWNING, AND HATCHING OF CARPS

6.1 Abiotic Factors

In general, the growth and development of a gamete depends on different environmental cues in nature. The primary environmental cues are often linked to the annual cycles of day length and temperature variations. For example, in case of *Cyprinus carpio* and many other cyprinids, gonadal maturation begins in late winter after the vernal equinox (21st March) when day length shows an increasing trend and as a result water temperature also registered an increasing trend. But in summer when temperature is high, cessation of spawning activity may occur due to the inhibitory effects of high water temperature. Thus, it is thought that both photoperiod and temperature are major environmental cues responsible for the mediation of the reproductive cycle in several species of fish that inhabit temperate latitudes. In temperate areas salmon generally breed at a

Induced Fish Breeding. DOI: http://dx.doi.org/10.1016/B978-0-12-801774-6.00006-7

temperature range of 8–10°C, i.e., during late autumn and early winter, but during autumn and winter when the temperature of river water may fall to 1–2°C, inhibition of oocyte growth and development may occur. Several environmental factors, such as light, temperature, and pH have a great influence on the gonadotrophic activity of the pituitary gland of fishes, and play an important role in stimulating the release of pituitary gonadotrophins, thus controlling the reproduction in fishes. Of these, light and temperature have been studied in detail by several investigators. Among other factors, rain, flood, sudden increase in water level, increased dissolved oxygen content, pH of water, repressive factors, such as petrichor, are also considered important for natural spawning of Indian Major Carps. Some investigators consider that the flow of water and the phases of moon also exert influence on the spawning behavior of fishes.

6.1.1 Light (Day Length)

Light appears to be an important factor in controlling the reproduction of fishes. Buliough (1941) reported that the reduction or total absence of light caused a delay in the maturation of the ovary in *Phoxinus laevis*. In *Salvelinus fontinalis*, functional maturity can be induced several months in advance of the normal time if they are exposed first to an increasing photoperiod, and later to a decreasing photoperiod (Hazard and Eddy, 1951), Henderson (1963) suggests that an accelerated light regime can hasten the time of functional maturity in trout. Harrington (1956, 1957) also suggests that long photoperiods initiate the gonadal activity and the fishes can spawn in advance of the natural time. On the other hand Shiraishi and Fukuda (1966) observed early maturation under short photoperiod and delayed maturation under long photoperiod. Verghese (1967) reported *Cirrhina reba* attains early maturity when exposed to longer photoperiods. According to Sanwal and Khanna (1972), both very long and very short photoperiods are unfavorable for maturation of ovaries in early stages in *Channa gachua*, and there is a delay in the appearance of yolk. But during vitellogenesis, a short photoperiod accelerates the formation of mature oocytes. All these studies clearly indicate that the influence of light in activating the productive cycle varies from species to species.

6.1.2 Temperature

The temperature also plays an important role in the maturation of gonads of fishes. Several authors (Hoar and Robertson, 1959; Ahsan, 1966) have shown that there is an optimum temperature for breeding of fishes, above and below which they may not reproduce. The maturity is suppressed at a low temperature in several species like *Fundulus*, *Tilapia*, and *Percafluviatillis*. On the other hand, Ahsan (1966) has shown that warm temperature stimulates maturation of gonads in several fishes and brings about spermiation. During induced breeding of Indian carps, spawning occurs at a temperature ranging from 24°C to 34°C. According to Chaudhuri (1968), the optimum water temperature is about 27°C. Bhowmick (1969) observed spawning in *C. reba*, by increasing water temperature from 25°C to 29°C at the time of injection, and then allowing the water to cool. Temperature appears to serve as a triggering factor for spawning in carps.

Besides light and temperature, several other factors, such as flood and rain, current of water, increased dissolved oxygen, shallow inundated spawning areas, are considered to be important ecological factors for inducing natural spawning of carps in rivers and bundh. Spawning could be induced successfully in carps on cloudy rainy days, especially after heavy showers. According to Chaudhuri (1968), low pH, low total alkalinity, abrupt rise of water level, and a water temperature range of 27–29°C are favorable for spawning. According to some, the rise of water temperature due to its passage through heated soils is of great significance, and without the rise in water temperature to a certain degree, no spawning is possible. It appears, therefore, that fresh rainwater and flooded condition provide primary fresh rainwater and flooded condition provide primary stimulus to spawning and sex-play.

6.1.3 Habitat and Repressive factors

The nature of the habitat, for example, the presence of stones, plants, suitable substratum to lay eggs, also serves as an important stimulus for the fish to breed. Swingle (1956) has suggested that the excretory matter of fish released into the water constitutes the "repressive factor," a hormone-like substance which inhibits reproduction in fishes. Hence, fishes which do not spawn when overcrowded, do so when transferred to fresh water. This has been confirmed in *Cyprinus carpio*, *Carassius auratus*, and *Tilapia mossambica* by Tang et al. (1963). It is suggested that spawning takes place when the repressive factor is sufficiently diluted by the flood water in bunds or ponds. According to Tang et al. (1963), ammonia itself may not be the repressive factor, but many other excretory substances may bring about inhibition of spawning in carps.

6.1.4 Petrichor

Lake (1967) suggests that the factor which stimulates the fish to spawn in natural environment is produced when rainwater comes into contact with dry soil, and has been identified as "Petrichor" obtained by steam distillation of silicate minerals and rocks plentifully available in the breeding habitat. Fishes possess a remarkable olfactory sense, and the odor of petrichor works as a stimulant for spawning. Nikolsky (1962) has suggested that certain steroid hormones secreted by the ripe male into the water, evoke the females to spawn followed by mating. One of the primary hormones identified is "copulin."

It has been already established that the release of the gonads stimulating hormones from the pituitary is regulated by the hypothalamus. After being stimulated by the said minerals and male hormone, neurosecretory material (GnRH) is transported from the hypothalamus to the hypophysis through the nerve fibers (axon) of the hypothalamico–hypophyseal axis. Seasonal changes in the activity of the hypothalamic nuclei have been correlated with the reproductive cycle of fishes, which suggests that the hypothalamus exercises primary control over the secretion of hormones from the pituitary gland.

6.1.5 Photoperiod and Seasonal Cycles

In contrast to this, if fish are maintained under experimental conditions of constant photoperiod (e.g., 12L:12D—12h light and 12h dark each day and temperature) the fish may still spawn at approximately yearly intervals. This means there is a strong autonomous component to the reproductive cycle, and under these conditions this endogenous rhythm will have a periodicity of about 1 year. In constant environment conditions, fishes may spawn at intervals that are approximately but are significantly different from 1 year. Thus the periodicity of the rhythm is only approximately 1 year, i.e., it is circannual.

As a result, many fish species rely on the seasonally changing cycle of day length to match their annual cycles of reproduction and these cyclic events will be mediated through a number of endocrine changes. The neuroendocrine system directly controls the different phases of general development and the maturation of gametes.

6.1.6 Dissolved Oxygen Concentration

For better hatching performance the optimum dissolved oxygen concentration should be 7–9ppm. At low oxygen levels, the embryonic development is retarded.

6.1.7 pH

For better hatching performance, the optimum pH level should be between 7.4 and 8.4.

6.1.8 Salinity

Good hatching results can be obtained if the water is free from salinity. Tolerable limits of salinity for hatching carp eggs are 10–100 mg/L of water, above which hatching is adversely affected.

6.1.9 Alkalinity

Good hatching results are obtained in waters having alkalinity of 150–250 ppm.

6.1.10 Silt

Deposition of silt on the fish eggs retards the embryonic development and destroys the eggs.

References

Ahsan, S.N., 1966. Some effects of temperature and lighton the cyclical changes in the spermatogenetic activity of the lake chub, in (Agassiz). Can. J. Zool. 44, 161–171.

Bhowmick, R.M., 1969. Economics of induced breeding of Indian major carps and Chinese carps—F.A.O./U.N.D.P. Regional Seminar on Induced Breeding of Cultivated Fishes, India. FR1/1BCF/20, p. 13.

Chaudhuri, H., 1968. Breeding and selection of cultivated warm water fishes in Asia and Far Dast. FAO Fish. Rep. 4 (44), 30–66.

Harrington, R.W., 1956. An experiment on the effects of contrasting daily photoperiods on gametogenesis and reproduction in the centrarchid fish, *Enneacanthus obesus* (Girard). J. Exp. Zool. 131, 203–223.

Harrington, R.W., 1957. Sexual photoperiodicity of the cyprinid fish, *Notropis bifrenatus* in relation to the phase of its annual reproductive cycle. J. Exp. Zool. 135, 529–553.

Hazard, T.P., Eddy, R.E., 1951. Modification of the sexual cycle brook trout (*Salvelinus fontinalis*) by control of light. Trans. Am. Fish. Soc. 80, 158–162.

Henderson, N.E., 1963. Influence of light and temperature on the reproductive cycle of the eastern brook trout, *Salvelinus fontinalis* (Michill). J. Fish. Res. Bd. Canada 20, 859–897.

Hoar, W.S., Robertson, G.B., 1959. Temperature resistance of goldfish maintained under controlled photoperiods. Can. J. Zool. 37, 419–428.

Lake, J.S., 1967. Rearing experiments with five species of Australian freshwater fishes. I. Inducement of spawning. Aust. J. Mar. Freshwater Res. 18, 137–153.

Nikolsky, G.V., 1962. On participation of geneticists in working out biological fishery problems. Vestn. Mosk. Univ. (Biol.) 6, 3–17.

Sanwal, R., Khanna, S.S., 1972. Seasonal changes in the testes of a freshwater fish, *Channa gachua*. Acta. Anat. 83, 139–148.

Shiraishi, Y., Fukuda, Y., 1966. The relation between the daylength and the maturation in four species of salmonid fish. Bull. Freshwater Fish. Res. Lab. 16, 103–111.

Swingle, H.S., 1956. A repressive factor controlling reproduction in fishes. Proceedings of The Pacific Science Congress 8, 865–871.

Tang, Y.P., Y.W. Hwang, Liu C.K., 1963. Preliminary report on injection of pituitary hormone to induce spawning of Chinese carps. Food and Agriculture Organization of the United Nations, Indo-Pacific Fisheries Council, Occasional Paper 63/14.

Verghese, P.U., 1967. Prolongation of spawning season in the carp *Cirrhina reha* (Ham.) by artificial light treatment. Curr. Sci. 36, 465–467.

CASE STUDIES

7

CURRENT STATUS OF HATCHERY OPERATIONS IN SOME LEADING SEED-PRODUCING STATES OF INDIA AND ITS IMPACT ON AQUACULTURE

CHAPTER OUTLINE

Induced Fish Breeding. DOI: http://dx.doi.org/10.1016/B978-0-12-801774-6.00007-9

7.1 Three Eastern States of India

Of the 29 states, West Bengal is one of the pioneering and most productive states, having an area of 88,551.0 sq. km for fish culture. Standardization of induced breeding technology, the fish farmers of Bengal primarily implemented this novel technology with subsequent establishment of eco-carp hatcheries and became pioneers in quality seed production to fulfill the total seed requirement of the country. In the freshwater sector commercial seed production started with Indian major carp from the late 1960s. Realizing the huge profit in from the short breeding season, peoples from diverse sectors started establishing hatcheries from 1960 onward. However, recently the breeding and culture of some exotic catfishes has been taken up by a considerable number of farmers on a commercial scale. Of these the Thai magur (*Clarias macrocephalus*), African magur (*Clarias gariepinus*), and Pangasid catfish (*Pangasius sutchi*) are the most important. Recently Pacu (*Piaractus brachypomus*), a native of the Amazon, has been introduced in the farming sector of Bengal. Already captive breeding and larval rearing of the introduced fish have been standardized (Chatterjee et al., 2009). They are fast-growing, very hardy, and unlike carp are able to withstand considerable variation in the environment. Attempts being initiated to standardize the breeding and rearing management of some of the indigenous varieties of endangered live fish and minor carps.

7.2 Categorization of Hatcheries

The hatcheries, based on their size and infrastructure, are generally categorized into small, medium, and large as depicted in Table 7.1. Major seed production activities, both at the start and end of the season, are controlled by the large hatcheries. Small and medium-sized hatchery owners take part during the midseason only due to the dire scarcity of brood fish and are subsequently deprived of getting a higher price (Gupta et al., 2000) as shown in Table 7.1.

7.3 Temporal Trends in Hatchery Establishment

Hatchery establishment started during 1961–70, boosted up to 1980–2000 with the successive introduction of alien species such as *C. gariepinus*, *C. macrocephalus*, *P. sutchi*, and *P. brachypomus*. As a result, most of the hatcheries established during this period were comprised of noncarp hatcheries, mainly *C. gariepinus* and in

Table 7.1 Categorization of Eco-Carp Hatchery

Criteria		Category	
	Small	Medium	Large
Farm area	0.70 ha	2–3 ha	3–5 ha
Actual brood-raising area	0.26 ha	1 ha	1.5 ha
Nursery space	0.25 ha	1 ha	1.5 ha
Total brood requirement	130 kg	1300 kg	2600 kg
Spawn production level (SPL)	5 million	50 million	80 million
Number of breeding operations	1/week	2/week	3/week
Diameter of spawning pool	3 m	3–4 m	4–5 m
Diameter of incubation pool	2 m	2.5 m	3 m
Number of spawning pools	1	1	1
Number of incubation pools	2	3	3–4

some cases *pangasid cat fish*. In Assam, during 1990–2000, initially 18 hatcheries were established and the number increased to 20 during 2000–10. The number has gone above 1000 since then. In Bihar, hatchery establishment started in 1990–2000 when only three hatcheries were established but the number is increasing almost daily.

Interesting to note that the innovative Bengal fish breeder started developing new hatchery design for the introduced alien species and sometime also found to improvise the existing one.

7.4 Ownership Status and Educational Profile of the Hatchery Owners

Ninety percent of the hatcheries studied came under single ownership, with only 3% under joint venture. Therefore, it is clear that entrepreneurship with single ownership control is preferred in all three states, possibly due to the infusion of technology primarily from Bengal. More than 67% of the fish breeders did not complete their school-level education, only 17 farmers were metric standard, 6 were higher secondary degree holders, and 5 fish breeders were graduates, 4 in commerce and 1 in the arts. It is striking to note that none were science graduate, despite the fact that they are practicing a scientific endeavor with fish seed production and hatchery management. Only one hatchery owner was an honors graduate in commerce, with a professional MBA degree. It might be concluded that the scientific knowledge bank through institutional training is rather poor among the fish breeders of all the states under study as far as their educational qualifications are concerned. This acts as a primary criterion to motivate illiterate fish breeders to adopt unscientific activities, which exerts a negative force against the principles of technology, i.e., production of quality seed in captivity.

7.5 Size of the Hatcheries and Their Modes of Operation

As per the criteria adopted for categorization of hatchery size (Table 7.1), most of the hatcheries come under the medium-sized category (21), followed by small (18) and large (10). It has been observed that more than 80% of hatcheries operate with Chinese-type eco-hatcheries, whereas 40% operate both with Chinese and another type, and only around 5% operate without a Chinese-type hatchery. The glass jar incubation system has been completely replaced by the Chinese hatchery system, due to its economic viability, easy operation, and the assertation of 80–90% fertilization and hatchability. Some fish breeders have developed innovative and economic modifications of their Chinese hatcheries for alien species, like *P. sutchi*, *C. gariepinus*, and *C. macrocephalus*, while some fish breeders developed modifications of the glass jar, which is not only economic but requires less space and effort with an assertion of maximum fertilization and hatching. This also focuses the innovative instinct of West Bengal breeders, not found in other states, in standardizing the propagation of any alien species.

7.6 Pond Status, Water Source, and Manpower

Fifty percent of the hatchery owners operate with their own ponds, whereas 6% of the breeders depend solely upon the leased ponds, 40% of farmers operate in ponds which they own as well as on a lease basis. More than half of the total fish breeders investigated absolutely depend upon the underground water source and the rest use both underground supplies as well as water from other sources, like rivers, ponds, etc.

7.7 Breeding Season and Price Variation

The season, as reported by the farmers, begins in March with the breeding of grass carp and continued till August or September in some cases. Usually breeding of Indian major carp (IMC) is undertaken from the month of April. The spawn price varies with the

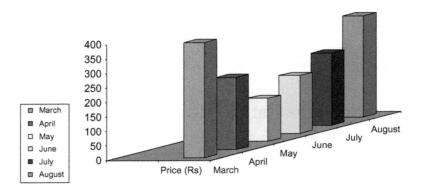

Figure 7.1 Spawn price variation with month.

season, during the early days of March–April (premonsoon) the price tends to be Rs. 400 per bati (beli in Assam) with a holding capacity of 30,000–50,000 spawn. The price slowly begins to decline to Rs. 150 per bati as the season approaches its peak during May–July (monsoon). In the later part of July and up to August there is again a price rise to the same level as (Fig. 7.1). Factors which primarily control spawn price include the availability of broodstock as well as the category of hatchery owners participating in a particular time of the breeding season.

7.8 Breeding Program

Initially, the fish breeders used to conduct breeding programs in breeding and hatching hapa. Afterwards with the development of Chinese-type eco-hatcheries, realizing the limitation of hapa breeding, the fish breeders of Bengal successfully adopted the same. This led them as the pioneer seed producers of the country who, initially, shouldered the responsibility to produce 70% of the seed requirement of the country. As the construction of eco-hatcheries requires huge initial investment (the cost of establishing Chinese hatchery) so the marginal and middle-category breeders could not afford such construction and used to conduct breeding programs conveniently in both Chinese hatcheries as well as in other available infrastructure:

1. Chinese-type eco-hatcheries: 80%
2. Both Chinese and others: 40%
3. Without Chinese hatcheries: 20%.

Realizing the huge water budget of breeding pool operation, more than 60% of fish breeders switch on to conducting breeding programs in an incubation pool, while some are of the opinion that because of

Figure 7.2 Breeding pool, a huge masonry structure, remains unutilized throughout the breeding season to reduce the water budget.

the sloping nature of the floor of the breeding pool toward the central orifice, most of the eggs after spawning concentrate in and around this orifice and reduce the fertilization rate. This is in contrast with the customary practice of conducting breeding in a separate breeding pool for each species to avoid indiscriminate hybridization. Only 12% of fish breeders still use conventional breeding pools and 6% undertake breeding in breeding hapa, 24% of farmers use a combination of these practices, both in breeding pools as well as in hapa. During the peak of the breeding season we have noticed breeding pools, large masonry structures, remain unattended (Fig. 7.2).

According to the site of operation of breeding the following categorization can be done:

a. Incubation pool: 60% (to avoid water requirement)
b. Breeding pool: 12%
c. Breeding hapa: 6 %
d. Breeding pool and hapa: 24%.

7.9 Brood Stock Management

As indicated in Table 7.1, the small farmers depend entirely on external sources for the collection of brood fish prior to the breeding program. The large and medium-sized farmers depend both on farm-raised and outside sources. Brooders are fed with mustard oil cake, semipolished rice, and grain and other ingredients depending upon the availability and cost at 200–250 kg/bigha per month at intensive stocking density throughout the year. Before the start of the breeding season there is a change in feeding and from February onwards the brooders are fed with specially prepared feed composed of groundnut oil cake (15 kg), mustard oil cake (10 kg), rice (20 kg), and cereals (2 kg) per week per bigha. This prebreeding shift in feeding schedule is to enhance maturity. The fish breeders, due to lack of proper scientific knowledge, face difficulties during rearing, sometimes due to overfeeding or due to lack of essential vitamins, minerals, and protein in the right ratio and combination in feed. Reports of protruded eyes and surfacing of broods during the afternoon due to supersaturation of oxygen are frequently reported by farmers. All these unscientific rearing practices delay maturation as well potency and promote defective gonadal development. This very negative and unscientific approach has already done significant damage to the seed production sector through the production of defective spawn and fry. Therefore, the sector initially blessed with the discovery of a novel technology, has now become a curse in disguise due to repeated misappropriation of the technology by fish breeders out of their greed to derive unaccountable profit.

7.10 Breeding Practice

7.10.1 Age and Size of Brooders

Table 7.2 depicts the average age and size of the brooders used by the hatchery operators of three states. In 60% of hatcheries the average age was 6.5–8.3 years and in 48% hatcheries was 0.53–5.29 kg. This shows extensive use of fish more than 5 years of age and size less than 0.53 kg as brood stock.

7.10.2 Sex Ratio

The average sex ratios considered for breeding in have been found to be 5:7 (M:F). Forty percent of the hatchery owners in the region have a sex ratio less than 1:1 (M:F) (Table 7.2). The breeders believe that one ripe and healthy male is sufficient to fertilize eggs of two or more females. Moreover, there is always a shortage of mature and healthy brooders in the hatcheries. That is why the breeders use the skewed sex ratio for achieving the targeted production without considering the future consequences.

7.10.3 Trend in Seed Production

Nearly 70% of the fish breeders operate their hatchery for production of carp (native and exotic) seeds along with either catfishes or with minor carps. It is interesting to note that there was no prawn

Table 7.2 Breeding Practices in Hatcheries of Three Leading Seed-Producing States of India, i.e., Bengal, Bihar, and Assam

Actors Involved in Fish Seed Production		Hatchery Owners and Fish Seed Producers		
		Hatcheries ($n = 25$)		
	Indicators	Variables	No. of Respondents	Percentage
Factors	Breeding age[a]	6.5–8.3 years[b]	15[c]	60
	Size	0.53–5.29 kg[b]	12[d]	48
	Sex ratio (skewed)	M:F = 5:7[b]	10[e]	40
	Multiple spawning	2 (Gc,Sc,Bh)	17	72
		3 (IMC,Mc)		
	Mix spawning	Multispecies	19	76

[a]Maximum breeding age = (minimum spawning age + no. of times a fish is spawned).
[b]Average.
[c](>5 years).
[d](<0.5 kg).
[e](<1:1), Gc, grass carp; Sc, silver carp; Bh, big head carp; Mc, minor carp.

hatchery in the area of the study. About 30% of hatcheries practice breeding and seed production of only exotic magur, *C. gariepinus*, and *C. macrocephalus* though it has been banned by the Government of West Bengal; because of its huge profitability and easy management practices, farmers are being lured to undertake seed production of this fish. This is similar for bighead carp (*Aristichthys nobilis*), for which more than 40% of the hatcheries successfully operate. Among the minor carp, bata, japani punti, and Java punti are taken up along with Indian major carp in 40% of the hatcheries. Seed production of Indian magur *Clarias batrachus* is totally absent within the purview of our study area except for one or two stray cases, although adult fish fetch a high price due to their high medicinal value and smaller spine. Two farmers practice culture and breeding of red rohu (a variety of *Labeo rohita*), this fish is subject to easy poaching because of its attractive color and surfacing behavior. Even up to a depth of 1.5 m they are visible from the water surface, which renders them prone to easy poaching as well as being taken by birds of prey. This fish, due to its dorsal and front beautiful red color (Fig. 7.3) is considered as a

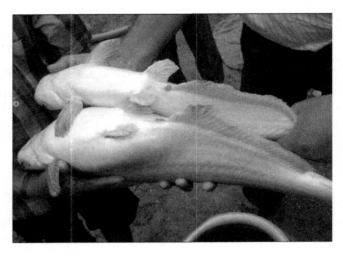

Figure 7.3 Male and female broods of *Pangus*.

most potent candidate species for culture and ornamental purposes, however, the fish breeders didn't show any interest for the same. The study revealed that the entire decision on present activities and future planning on hatchery operations in three states is controlled by the large hatchery owners, money lenders, middlemen, and also agents and dealers of big business sectors dealing with fertilizers, feed, medicines, drugs, probiotics, and a number of unauthorized products.

1. Carp (native and exotic) + catfish + minor Carp: 70% (1980)
2. Only exotic magur: 30%
3. Bighead: 40%
4. Pangasius: 15%
5. Red rohu: 2%
6. Pacu: 10%.

7.10.4 Inducing Agent, Mode of Breeding, and Restocking

More than 80% of the fish breeders of West Bengal and Assam use pituitary extract, due to lower cost, easy availability, and possibly due to its organic nature. The quality of the gland is assured by its physical appearance, mainly color, shape, and elasticity. Freshly preserved gland is more elastic and brownish. On the contrary, it is soft and pale yellow when it is not properly preserved. As per farmers' experience, during windy and rainy days the dose is reduced. Also, a sandy soil bed requires a lower dose of induction. Other than pituitary, only

Table 7.3 Mode of Breeding, Inducing Agent, and Restocking as Observed in Three Leading Seed-Producing States of India, i.e., Bengal, Bihar, and Assam

Actors Involved in Fish Seed Production			Hatchery Owners and Fish Seed Producers of Three States	
		Hatcheries (n) = 25		
	Indicators		No. of Respondents	Percentage
Factors	Natural breeding		18	72
	Stripping		8	32
	Restocking		19	76
	Inducing agent	CPE	12	48
		Ovaprim	20	80

a few farmers use ovaprim, ovatide, WOVA-FH, etc., because of their high price and the difficulty in calculating the dose and also because of the long-term dependency, and belief that synthetic products are less active than pituitary extract. On the contrary, Assam breeders use ovaprim in more than 80% of breeding practices (Table 7.3).

7.10.5 Rate of Water Flow in Breeding Pool and Incubation Pool

1. Breeding pool: The prescribed flow rate is 30–50 L/minute.
2. Incubation pool: The prescribed rate ranges from 45 to 60 L/minute. The farmers adjust the rate in such a way that it would not allow the developing fertilized eggs to settle at the bottom, instead the eggs undergo a cyclic downward and upward movement until hatching, unaware of the recommended rate. The fish breeders are so skilled that they maintain prespawning, spawning, and postspawning flow rates simply by observing the successive developments in the hatching pool.

The flow rates maintained in the breeding and hatching pools are recorded in Table 7.4. The initial flow rates, which begins 2–3 hours before the actual spawning occurs in the spawning pool, were recorded as 0.25 ± 0.02 (min) and 0.52 ± 0.056 (max). The final flow

Table 7.4 Flow Rate in Breeding and Hatching Pool

Breeding Pool (m/s)		Hatching Pool (m/s)		
Initial	Final	Initial	Mid	Final
Mean±SD		Mean±SD		
0.37 ± 0.03	0.22 ± 0.066	0.18 ± 0.03	–	0.13 ± 0.02
0.33 ± 0.085	0.18 ± 0.040	0.15 ± 0.015	0.14 ± 0.015	0.14 ± 0.015
0.29 ± 0.036	0.17 ± 0.041	0.2 ± 0.032	0.13 ± 0.015	–
0.32 ± 0.03	–	0.19 ± 0.026		0.31 ± 0.02
0.52 ± 0.056	0.29 ± 0.025	0.24 ± 0.035	0.16 ± 0.020	0.21 ± 0.060
0.25 ± 0.02	0.17 ± 0.015	0.30 ± 0.025	0.21 ± 0.015	0.30 ± 0.025

rates recorded were 0.17 ± 0.015 (min) and 0.29 ± 0.025 (max). Similarly, the initial flow rates observed for the first 12 hours in the hatching pool were 0.18 ± 0.03 (min) and 0.31 ± 0.02 (max). The mid-flow rates observed in the hatching pool for another 6 hours were 0.13 ± 0.015 (min) and 0.21 ± 0.015 (max). While the final flow rates for the rest of the operation were 0.13 ± 0.02 (min) and 0.31 ± 0.02. ANOVA confirmed a highly significant difference between the initial and the final flow rates maintained in the breeding pool by the farmers ($F = 22.51$, $p < 0.05$) followed by t-test ($t = 4.96$, $p < 0.05$) also showed significant differences. A significant difference ($F = 4.76$, $p < 0.05$) was also noticed in the flow rates of the hatching pool during initial (first 12 hours), mid (next 6 hours), and final (rest period) phases of the breeding period. The results suggest that highly significant differences between the initial and the final flow rates are maintained in the breeding pools (Fig. 7.4). Similar significant differences in the flow rates were observed in the hatching pools (Fig. 7.5).

7.10.6 Pituitary Protocol Used in Hatcheries of Three States

The average pituitary protocol used by different hatcheries of Barak Valley during premonsoon, monsoon, and postmonsoon are presented in Table 7.5. It was observed that the dose of CPE was higher during premonsoon and postmonsoon, while the dose in the monsoon season was comparatively less. This may be due to brooders not attaining full maturity during the premonsoon period, or congenial environment not available, etc. During postmonsoon, the

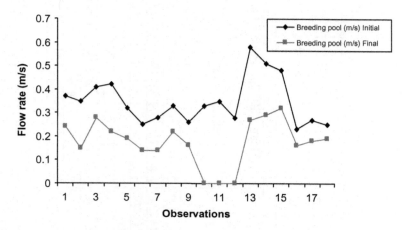

Figure 7.4 Flow rate in breeding pool.

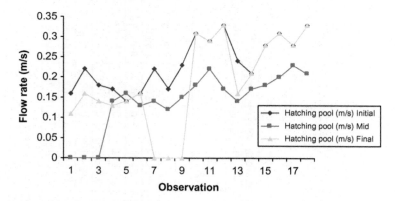

Figure 7.5 Flow rate in hatching pool.

brooders used are already spent twice and are by then on the recession stage. Catla preferred a slightly higher dose compared to rohu and mrigal throughout the season as reported. The male dose does not vary much irrespective of breeding season.

7.10.7 Ovaprim Protocol Used in Hatcheries

Table 7.5 depicts the average ovaprim protocol used during premonsoon, monsoon, and postmonsoon seasons. It was observed that the dose of ovaprim did not vary much with the breeding season as was seen with the pituitary gland. It is evident from the table that no significant difference in ovaprim dose for males and females of rohu, mrigal, and catla during premonsoon, monsoon, and postmonsoon seasons was observed.

Table 7.5 Comparative Study of Spawn Yield with CPE and Ovaprim (N = 25)

	Carp Pituitary Extract (mg/kg) (Mean ± SD)						(Mean ± SD) Spawn Yield (lakh/kg)		
	Premonsoon		Monsoon		Postmonsoon		Premonsoon	Monsoon	Postmonsoon
	F	M	F	M	F	M			
	1	2	3	4	5	6	7	8	9
Rohu & Mrigal	12.66 ± 0.57	4 ± 1	9.33 ± 0.57	2.33 ± 0.57	13.66 ± 1.52	3.66 ± 0.57	0.5 ± 0.1	0.7 ± 0.1	0.23 ± 0.05
	13.33 ± 0.57	3.66 ± 0.57	9.33 ± 1.15	3 ± 0	13.33 ± 3.05	3.33 ± 0.57	0.63 ± 0.23	0.7 ± 0.2	0.33 ± 0.05
	13.33 ± 1.15	4.33 ± 0.57	9.66 ± 0.57	3.33 ± 1.52	14.33 ± 0.57	4.33 ± 0.57	0.56 ± 0.057	0.63 ± 0.15	0.35 ± 0.1
Catla*	14.33 ± 1.15	3.33 ± 0.57	11.66 ± 0.57	3 ± 0	14.66 ± 1.15	3.66 ± 0.57	–	–	–
	14.33 ± 0.57	4.33 ± 0.57	11.33 ± 1.15	3 ± 1	14 ± 1.73	3.66 ± 0.57	–	–	–
	13.33 ± 0.57	5.33 ± 0.57	12.66 ± 1.15	4.33 ± 0.57	14.33 ± 0.57	5.33 ± 0.57	–	–	–
Ec (Bh Sc)	–	–	13.66 ± 0.57	4.33 ± 0.57	15 ± 1	4.33 ± 1.15	–	0.6 ± 0.1	0.3 ± 0.1
	–	–	13.33 ± 0.57	4.33 ± 0.57	14.5 ± 0.57	4.66 ± 0.57	–	0.66 ± 0.05	0.26 ± 0.05
	–	–	12.66 ± 0.57	3.33 ± 0.57	14 ± 0	4 ± 0	–	0.63 ± 0.115	0.36 ± 0.05

	Ovaprim Dose (mL/kg) (Mean ± SD)						(Mean ± SD) Spawn Yield (lakh/kg)		
	Premonsoon		Monsoon		Postmonsoon		Premonsoon	Monsoon	Postmonsoon
	F	M	F	M	F	M			
	10	11	12	13	14	15	16	17	18
Rohu & Mrigal	0.46 ± 0.05	0.15 ± 0.05	0.4 ± 0.1	0.1 ± 0	0.46 ± 0.05	0.11 ± 0.02	0.7 ± 0.1	1.13 ± 0.40	0.4 ± 0.1
	0.46 ± 0.05	0.13 ± 0.02	0.33 ± 0.05	0.13 ± 0.05	0.36 ± 0.11	0.13 ± 0.02	0.9 ± 0.26	1.06 ± 0.37	0.5 ± 0.1
	0.5 ± 0	0.2 ± 0	0.3 ± 0	0.1 ± 0	0.5 ± 0	0.2 ± 0	0.86 ± 0.28	0.96 ± 0.20	0.4 ± 0.1
Catla*	0.43 ± 0.05	0.16 ± 0.02	0.4 ± 0	0.1 ± 0	0.43 ± 0.05	0.11 ± 0.02	–	–	–
	0.5 ± 0	0.18 ± 0.02	0.4 ± 0.0	0.11	0.5 ± 0.0	0.133	–	–	–
	0.43 ± 0.05	0.13 ± 0.05	0.33 ± 0.05	0.11 ± 0.8	0.4 ± 0.1	0.11 ± 0.02	–	–	–
Ec (Bh Sc)	–	–	0.66 ± 0.05	0.16 ± 0.05	0.56 ± 0.05	0.15 ± 0.05	–	0.73 ± 0.057	0.36 ± 0.152
	–	–	0.5 ± 0	0.1 ± 0.0	0.6 ± 0.1	0.13 ± 0.02	–	0.76 ± 0.05	0.33 ± 0.05
	–	–	0.43 ± 0.05	0.11 ± 0.02	0.56 ± 0.11	0.18 ± 0.02	–	0.56 ± 0.05	0.36 ± 0.05

CPE = Carp Pituitary Extract; Ec = Exotic carp; Bh = Bighead; Sc = Silver carp

7.10.8 Study of the Effect of Carp Pituitary Extract (CPE) and Ovaprim on Latency Period, Effective Spawning Time, and Incubation Period

The study area was divided into six groups, G1, G2, G3, G4, G5, and G6; each group consisting of three hatcheries on the basis of use of inducing agent (CPE or ovaprim). Thus, a total of nine hatcheries for each inducing agent (CPE or ovaprim) and a total of 18 hatcheries were observed. The mode of fish breeding operation was closely observed to study the latency period, incubation period, and effective spawning time with the use of CPE and ovaprim. The weight of the fish, sex ratio, and dose of CPE and ovaprim administration were recorded.

The study was purely based on the method adopted by the farmer to breed the fish in the hatcheries for production of carp seeds. There was no scientifically based experimental set up for the study, it was purely a field study undertaken by the researcher. Male and female IMC were selected based on external morphological characteristics and experience. Pituitary extracts prepared by the farmer were injected intramuscularly in the caudal peduncle region above the lateral line to the female (initial dose) and after 5 hours males and females (resolving dose) were injected and released into the spawning pool (Figs 7.6–7.8).

The brooders started showing aggressiveness, and after a period of 5–6 hours spawning occurred, which continued for another 3 hours. Eggs were hatched out after a period of 16–18 hours. The time between administration of hormone and initiation of spawning (latency period); the actual time between initiation of spawning and completion of spawning (effective spawning time); and the time between fertilization and hatching of egg (incubation period) were recorded. The mode of fish breeding operation was observed to study the latency period, incubation period, and effective spawning time of IMC with the application of carp pituitary extract and ovaprim. The latency period, effective spawning time and incubation time, weight of the fish, sex ratio, and dose of ovaprim and CPE administration are recorded in Table 7.6. The latency period recorded with CPE is 10.6 ± 1.15 to 11.33 ± 1.15 hours and 7.83 ± 0.28 to 8.66 ± 0.57 hours with ovaprim. The effective spawning time was 3.66 ± 0.57 to 4.33 ± 0.57 hours with CPE and 1.83 ± 0.28 to 2.25 ± 1.06 hours with ovaprim. The incubation time varied between 17.6 ± 0.57 to 18.0 ± 2 hours with CPE and 13.33 ± 1.57 to 15.33 ± 0.57 with ovaprim. ANOVA confirmed that ovaprim exerted a significant effect ($p < 0.05$) on latency period, effective spawning time, and incubation time when compared to pituitary gland extract (Figs 7.9–7.11).

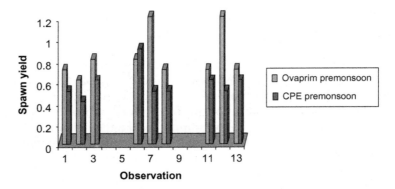

Figure 7.6 Effect of CPE and ovaprim on spawn yield during premonsoon.

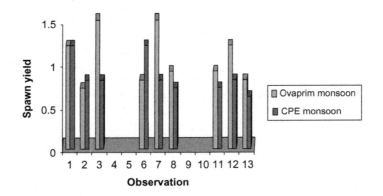

Figure 7.7 Effect of CPE and ovaprim on spawn yield during monsoon.

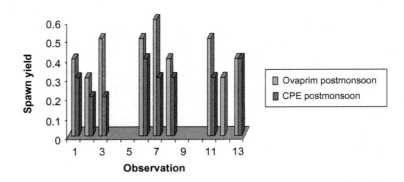

Figure 7.8 Effect of CPE and ovaprim on spawn yield during postmonsoon.

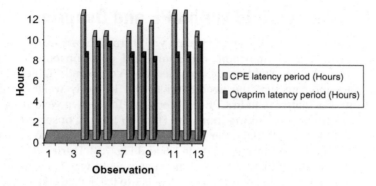

Figure 7.9 Effect of CPE and ovaprim on the latency period.

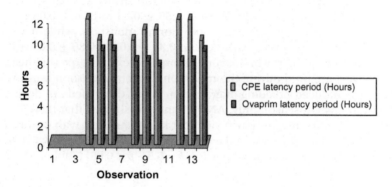

Figure 7.10 Effect of CPE and ovaprim on effective spawning time.

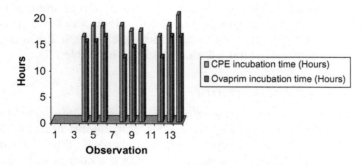

Figure 7.11 Effect of CPE and ovaprim on incubation time.

7.11 Spawn Yield With CPE and Ovaprim

The spawn yield per kg body weight with carp pituitary extract and ovaprim during premonsoon, monsoon, and postmonsoon in the hatcheries of three states were recorded and compared (Table 7.5). The data indicate that the average yield per kg body weight with ovaprim was higher than in CPE-injected fish. The spawn yield per kg body weight for IMC during monsoon (June to July) as observed was 0.63 ± 0.15 to 0.7 ± 0.1 lakh per kg with CPE and 0.96 ± 0.20 to 1.13 ± 0.40 lakh per kg with ovaprim (Table 7.5 and Fig. 7.6). During the premonsoon season (March to April) the spawn yield was 0.5 ± 0.1 to 0.63 ± 0.23 lakh per kg with PG and 0.7 ± 0.1 to 0.9 ± 0.26 lakh per kg with ovaprim (Table 7.5 and Fig. 7.6). The lowest quantity of spawn yield as observed was 0.23 ± 0.05 to 0.35 ± 0.1 lakh per kg with PG and 0.4 ± 0.1 to 0.5 ± 0.1 lakh per kg with ovaprim (Table 7.6 and Fig. 7.8) during postmonsoon season (August to September). Similarly the spawn yield for exotic carp (big head and silver) was 0.6 ± 0.1 to 0.66 ± 0.05 lakh per kg with PG and 0.56 ± 0.05 to 0.76 ± 0.05 lakh per kg with ovaprim, was maximum during monsoon, while it was 0.26 ± 0.05 to 0.36 ± 0.05 with PG and 0.33 ± 0.05 to 0.36 ± 0.05 with ovaprim during the postmonsoon season. The exotic carps (big head and silver) are bred during the monsoon and postmonsoon seasons, while the induced breeding of grass carp is mostly carried out during the premonsoon season only. ANOVA analysis showed that there is a significant difference in spawn yield per kg body weight with ovaprim and CPE, i.e., ovaprim exerted a significant difference on spawn yield per kg body weight compared to CPE ($F = 5.51$, $p < 0.05$) (Table 7.6).

7.12 Brood Stock Management

7.12.1 Species and Stocking Density

Eighty percent of the hatcheries maintain a multispecies brood stock composed of IMC, minor carp, and exotic carp of different age groups and sizes with high stocking density, along with banned species. A stocking density of more than 1500 kg/ha is 72%. The average stocking density as maintained was 2315.79 kg/ha (Table 7.7).

7.12.2 Feed and Fertilizer

The majority of the hatchery managers do not have proper feeding schedules for the brooders and nurseries. The majority of hatcheries (80%) provide feed and fertilizer on a monthly basis (Table 7.7). Just before the start of the breeding season there is a change in feeding schedule. From February onward the brooders are fed with mustard

Table 7.6 Effect of CPE and Ovaprim on Latency Period, Effective Spawning Time, and Incubation Time ($N = 18$)

Hormone	Species	Weight (kg)				Dose			Latency Period (h)	Effective Spawning Time (h)	Incubation time (h)
		F	M	Sex Ratio (M:F)	Total Weight	Female 1st	Female 2nd	Male			
CPE	Indian	0.81–1.9	0.5–1.35	5:7	32.66 ± 8.73	3.33 ± 0.57	6 ± 0	2.33 ± 0.57	10.6 ± 1.15	4.33 ± 0.57	17.33 ± 1.15
	Major	0.8–1.96	0.91–1.2		39 ± 11.53	3.5 ± 0.70	5.5 ± 0.7	3 ± 0	10.66 ± 0.57	3.66 ± 0.57	17.33 ± 0.57
	Carp	0.78–2.1	0.85–1.6		40 ± 4.24	3 ± 0	6 ± 0	2 ± 0	11.33 ± 1.15	3.66 ± 0.57	18 ± 2
		Mean	Mean	Mean	Mean ± SD						
Ovaprim		0.8–.23	0.85–1.5	5:7	57.5 ± 3.53	0.43 ± 0.57	–	0.1 ± 0	8.66 ± 0.57	1.83 ± 0.28	15.33 ± 0.57
		0.8–.96	0.93–1.2		28 ± 8.48	0.3 ± 0	–	0.1 ± 0	7.83 ± 0.28	2.0 ± 0.70	13.33 ± 1.15
		0.72–2.2	0.5–0.8		40 ± 0	0.5 ± 0	–	0.15 ± 0	8.33 ± 0.57	2.25 ± 1.06	14.66 ± 2.23

Table 7.7 Brood Stock Management Practice in Three States

Actors Involved in Fish Seed Production		Hatchery Owners and Fish Seed Producers		
		Hatcheries (n) = 25		
	Indicators	Variables	No. of Respondents	Percentage
Factors	Species	Multispecies[a]	20	80
	Stocking density	2315.79 kg[b]	18[c]	72
	Feed and fertilizer	Monthly	20	80
	Infrastructure	Insufficient	21	84
	Source of brood fish	Own/others	18	72
	Culling	Deformed/diseased/ overage/etc.	22	88
	Pedigree	–	20	80

[a](>6 species).
[b]Average.
[c](>1500) kg.

oil cakes (10 kg), rice polish (10 kg), and broken rice (5 kg) per week per 0.13 hectare pond. Just after the onset of the breeding season the feed is given regularly to spent brooders to regain quick maturation. Multiple spawning is a regular practice. For preparation of the brood stock pond, 18 kg (two bags) of lime (CaO) and 0.1 MT of raw cow dung per 0.13 hectare pond are applied.

7.12.3 Infrastructure for Brood Stock Raising

Considering the formula proposed by Thomas et al. (2003) 84% of the ponds don't possess sufficient water area for raising brood stock and this acts as an initial source for developing several negative genetic consequences as indicated in Chapter 5.

$$ABRP = BR \times 1/SD,$$

where ABRP = area of brood fish production pond; BR = weight of brood fish expressed in kilograms; SD = stocking density of brood fish per hectare.

7.12.4 Source and Pedigree of Brood Stock

Seventy-two percent of the hatchery owners, due to dire want of brooders, cannot continue sustained levels of spawn production during the premonsoon and even in monsoon seasons (Table 7.7). The small farmers depend entirely on outside sources for collection of brood fish prior to the breeding program. While 62% (average) of breeders own their own pond, the other 38% (average) use an outside source, and a share basis is prevalent among the farmers.

The breeders in the hatcheries (80%) visited during the study are least bothered to know the pedigree of fish stock they have for the breeding program to be carried out in their farms (Table 7.7). Not a single farm was found to maintain a pedigree record of the brood stock they have been maintaining. The pedigree record should be maintained to avoid the mating of close relatives. Hatchery operators should have detailed information on the pedigree of their brood stock. Cultured populations should be identified by using a proper marking system. Females and males should be from two different lines.

7.12.5 Culling

The practice of eliminating or culling of fish from the breeding program is a necessity of modern breeding programs. The culling of fish could be based on criteria including the phenotype of fish, such as growth, disease, deformity, age, size, and most catchable fish rather than the least catchable during harvesting. No such activity was adopted by any of the fish seed producers of the study area, instead one of the breeders was seen injecting a grass carp fish to induce spawning with a tumor-like lump on the skull region of the fish (Fig. 7.12, Plate D). According to the information provided by him, he had been breeding the same fish for the last couple of years. Table 7.7 confirms that the majority of the breeders have no idea about the importance of culling of fish from the breeding program.

7.12.6 Recommendation to Fish Hatcheries to Incorporate Genetics

7.12.6.1 Management Plans

This study demonstrates the utility of using a genetic management plan to maintain a captive fish population. Conservation breeding programs have their origins in zoos, in part, because these institutions manage much smaller populations that are easier to manage than the hundreds to thousands of fish that hatcheries traditionally manage each generation. The implementation of a genetic management plan might be more difficult and cumbersome for fish

Figure 7.12 Tumor at the junction of head and body of grass carp are in use for successive breeding program.

hatcheries; however, these institutions generally release large numbers of fish back into wild populations and have been documented to negatively impact the genetic diversity and effective population size of wild populations (Ryman and Laikre, 1991). Thus, it is important that fish hatcheries implement genetic management plans to prevent detrimental changes to supplemented wild populations.

Because the adoption and implementation of a genetic management plan in a large fish hatchery is costly and labor-intensive, genetic management plans can be tailored specifically to each hatchery. Efforts to maintain genetic diversity and prevent genetic responses to captive breeding are important to the success of both conservation and population supplementation. As a result, it is important for hatchery managers to determine what is cost-effective and physically possible to accomplish with a genetic management plan in a hatchery setting. A genetic management plan that includes analyses to attempt to equalize family contributions and allow for breeding schemes that minimize kinship may preserve the genetic integrity of the captive population. This may be accomplished using molecular markers, by keeping family groups in separate tanks, or by tagging individuals. Procedures not based on molecular markers may be relatively inexpensive and may still allow for the implementation of a simpler genetic management plan for those hatcheries where using molecular data to reconstruct the pedigree is not feasible. To reduce genetic adaptation to captivity in other hatcheries, fish may be exposed to simulated natural conditions in tanks or use an open flow-through water system that exposes captive fish to conditions in their native habitat. Regardless of the complexity or limitations of the genetic management plan, the following guidelines are recommended:

1. Pedigree analysis of the captive population should be conducted to allow equalization of family sizes and allow for a breeding scheme that minimizes kinship in the population.
2. Wild individuals should be periodically incorporated into the captive population to allow gene flow between the wild and captive populations in order to minimize genetic drift in the captive population, if possible.
3. Genetic adaptation to captivity should be minimized through exposure to naturalistic conditions.

The results of this study suggest that fish hatcheries utilizing genetic management plans designed to minimize mk can preserve genetic diversity in captive populations. Incorporation of wild fish into each generation is also an important component in the success of these programs; however, this is not feasible in highly endangered or extirpated populations, indicating the increased need for careful genetic management. Continued genetic management of the delta smelt captive breeding program may preserve this species should it become extinct in the wild and will serve as a model for fish hatcheries adopting hatchery genetic management plans.

7.13 Fish Disease

The most commonly occurring fish diseases in hatcheries are indicated in Table 7.8. The spawn and fries at a certain interval suffer from

Table 7.8 Factors Responsible for Inbreeding and Genetic Drift ($N = 25$)

Practice	No. of Farmers			Percentage	
	Yes	No	N/A	Yes	No
Partial replacement of brood stock (>3 years)	5	14	6	20	56
Effective breeding no. (<263)	–	19	6	–	76
Brood stock raised from same offspring (negative selection)	17	–	8	68	–
Skew sex ratio	10	9	6	40	36
Varied potency	12	5	8	48	20
Random mating instead of pedigreed mating	19	–	6	76	–
Brood size (>5 kg)	9	8	8	36	32

the outbreak of one or a mixture of uncontrollable diseases, killing lakhs of spawns and fries within a short time and imparting a heavy loss to the breeders. There are reports of the use of lime + $KMnO_4$ + salt to control diseases. Both in the cases of stripping and normal breeding the spawns are affected with crumpling disease in which the affected spawns assume a comma shape. This may cause mortality of up to 100%. There is report of sudden collapse of spawn and loss of the entire crop in the hatching pool or just after transfer of spawn (60–72 hours old) to the spawnery or hapa.

7.14 Deformities

All the deformities as indicated in Chapter 2 are observed in all three states.

7.15 Inbreeding

As mentioned elsewhere, inbreeding is a common phenomenon and the possible factors are indicated in Table 7.7.

7.16 Genetic Drift

Almost all the hatcheries with limited numbers of brood stock are prone to genetic erosion which may have accrued through inbreeding, genetic drift, and bottleneck effect in the hatchery populations.

7.17 Hybridization

About 10–15% of the IMC hatcheries purposefully hybridize IMC species, particularly among rohu, catla, mrigal, calbasu, and gonius and also between silver carp, bighead, and bata with mrigal following reciprocal crossing methods. Inclusion of hybrid fish into culture practices and escape into natural water bodies is now posing a threat to the very existence of prized fish of India.

7.18 Impact on Wild Fish

The researcher cannot comment on the entry of genetically eroded or hybridized fish into the wild since any collection of samples from the landing center could be taken up during the study period because of the ban imposed on fishing in natural water bodies from April to July. However, the fishermen of Assam have reported that there is ample evidence of hybridized fish sneaking into the wild territory since fish caught from the beels, haors, and anuas comprise hybridized fish. Francoise Ratz, a fishery expert (World Bank Preparatory Mission, 1999), in his report to the government of

Assam stated that several hybridized fish caught in the lower reaches of Brahmaputra in Bangladesh have been confirmed by Bangladesh Agriculture University (BAU), Mymensingh (Bangladesh). The fishermen communities of the area have reported that there is ample evidence of hybridized fish sneaking into the wild territories since fish caught from the beels, haors, and anuas comprise hybridized fish.

There is a scientific consensus that both intentional and unintentional release of cultivated fish pose serious risks to the genetic identity of the recipient populations (Ruzzante et al., 2004; Hansen et al., 2001). Unintentional selection in hatcheries may also cause changes in quantitative traits of the reared fish, such as increased growth rates and early maturation. The stocks of farmed and native fish might also have different historical backgrounds, and thereby potentially also harbor unique and different genetic features. Mixing between farm-raised and natural fish might therefore result in alterations of the wild genetic structure and in the breakdown of locally adapted gene complexes (Kallio-Nyberg and Koljonen, 1997; Fleming et al., 2000; McGinnity et al., 2003).

The adaptability of any hybrid to the environment is determined by genetic introgression, i.e., flow of genes from the gene pool of one species to another. If introgression is <0.1% it may help in increasing capability of adoption against natural selection. A large amount of gene flow may disrupt the adoptive gene complexes, which have evolved overnight to permit a species to effectively use its particular environmental niche, so while releasing any hybrid to nature, care should be taken to evaluate it properly (Thomas et al., 2003). Initially the breeding season for inducing fish in captivity was restricted from June to August. However, farmers have managed to extend this period by 3 months, i.e., from April until September. In case of magur (*C. batrachus*) sometimes farmers undertake breeding in winter by using comparatively cooler bottom water. This obviously indicates the farmers' skill of innovation in breeding.

7.19 Innovative Technology Developed by Scientists and Fish Breeders of Bengal by Modifying Glass Jar Hatcheries

7.19.1 Thums Up Bottle as an Economic Hatchery for Hatching of *Cirrhinus reba* (Hamilton, 1822) Eggs in Captivity

Induced spawning of freshwater *C. reba* (locally known as raikor bata) was carried out in captivity using ovaprim (sGnRH + domperidone). The optimum doses of ovaprim (0.3 mL/kg body weight for males and 0.5 mL/kg for females) were standardized based on three experiments, namely, fecundity (relative fecundity), response time (hours), and fertilization rate at different doses. Maximum no. of eggs is "9120" when measured through Gravimetric methods. Again maximum fecundity (9120) obtained at 0.5 mL/kg of body weight following batch fecundity for each female is calculated from the product of the number of hydrated oocytes (eggs) per unit weight in the tissue sample and the ovary weight (left and right sides combined). Hatching was conducted in a specially designed inverted 2-L Thums Up bottle following the mechanism of a glass jar hatchery and the spawns are reared at nursery ponds by placing the hatchlings in the hapa. This innovative and cost-effective technology will certainly add a new era to the seed production of *C. reba* and some other indigenous fish of the freshwater sector.

C. reba (Hamilton, 1822), commonly known as kharkebata or raikhor bata, is a widely distributed endemic freshwater minor carp of India, Bangladesh, Pakistan, and Nepal. In India, it is a very common species of rivers and ponds in the Gangetic belt of the northern region and also in the river Cauvery in the south. In West Bengal, especially in the northern part, it fetches considerable market demand for its consumer preference due to its good taste as well as soft and less bony flesh (Chondar, 1999). The initial growth rate is faster than catla and due to its compatibility the fish can be considered as a major candidate species for polyculture with Indian major carp (Job, 1944). In addition, the fish is adorned with an attractive silvery luster along with beautiful hexagonal scales and bluish longitudinal bands above the lateral line, which offers the fish an ornamental value. All these qualities allure farmers to consider the fish as a potential candidate species for diversified farming. Nonavailability of quality seed, due to want of any standardized procedure for breeding in captivity, means the population of such an important item for culture is decreasing rapidly. *C. reba* is a perennial breeder while the single spawning period restricts it the southwest monsoon extending from June to August in West Bengal. The fish does not spawn in captivity despite attaining full maturity. Chondar (1999) reported responsive results in captivity following induction by pituitary gland extract, even at a small dose. This study was conducted to standardize the captive breeding of the fish, using the synthetic inducing agent ovaprim, followed by stripping. Subsequently, hatching of the fertilized eggs was conducted using low-cost Thums Up bottles as a substitute for glass jar hatcheries.

7.19.1.1 Systematic Position

Phylum: Chordata
Class: Actinopterygii (ray-finned fish)

Order: Cypriniformes (carp)
Family: Cyprinidae (carp and minnows)
Genus: *Cirrhina*
Species: *C. reba*

7.19.1.2 Habitat

C. reba is found in large streams, rivers, tanks, lakes, and reservoirs (Menon, 1999; Talwar and Jhingran, 1991). It attains a total length of 30 cm (Menon, 1999) and breeds once a year in the flooded shallows regions of rivers during June to early September (Gupta, 1975). The fish feeds on plankton and detritus (Talwar and Jhingran, 1991). No information is available on the definite habitat trends of the fish and it is found in all the rivers and clear streams of Bangladesh (Bhuiyan, 1964a,b) including tanks, canals, ponds, beels, and inundated fields (Talwar and Jhingran, 1991; Rahman, 1989).

7.19.1.3 Food and Feeding

It feeds mainly on plankton and detritus (Talwar and Jhingran, 1991) but also on mud, vegetables, crustaceans, and insect larvae (Bhuiyan, 1964a,b). Feed ingredients are mainly composed of algae (10%), higher plants (70%), protozoa (5%), crustaceans (10%), and mud and sands (5%) (Mookherjee et al., 1946).

7.19.1.4 Collection of Brood Fishes and Their Acclimatization

Disease-free, healthy, and gravid males (average length 15 cm and weight 135 g) and females (average length 17 cm and weight 150 g) were collected from local fish ponds in June and were acclimatized for 48 hours, prior to breeding operation in a glass aquarium (120 cm × 45 cm × 45 cm) filled with unchlorinated tap water (pH 7.2 + 0.15, DO 5.72 + 0.80 mg/L, CO_2 10.16 + 2.3 mg/L, total alkalinity 146 + 6.13 mg/L as $CaCO_3$ and hardness 107 + 8.2 mg/L as $CaCO_3$) with aeration facilities. During acclimatization, male and female brooders were kept in separate aquaria side by side to enhance the desire for mating and no feed was applied. Before release into aquaria the fish were treated with 1% $KMnO_4$ solution for 5 minutes to ensure they were pathogen-free (Fig. 7.13).

7.19.1.5 Selection of Brooders

As in carp, gravid males are identified by the dorsal roughness of pectoral fins, stout abdomen with elongated, introverted, and whitish vent. On slight pressure on the abdomen prior to vent milt oozes out. The scales on the flanks, nape, and anterior–dorsal side are rough and sandy in texture. Fully gravid female brooders, on the other

Figure 7.13 Brood fish of raikhor bata, *C. reba.*

Figure 7.14 Mature male raikhor bata, *C. reba.*

hand, are characterized by tender bulging abdomen with extrovert, fleshy, round, and pinkish vent. On slight pressure eggs comes out one by one. Pectoral scales are smooth to touch on the dorsal side (Figs 7.14–7.16).

7.19.1.6 Breeding Operation

They generally breed at the beginning of summer which extends from June to September, when the temperature is high and rainfall is

Figure 7.15 Mature female *C. reba*.

Figure 7. 16 Stripped eggs in tray.

excessive, indicating a definitive role for temperature and rainfall on natural spawning (Bhuiyan, 1964a,b). The fish breeds along the shallow inundated region along the river (Talwar and Jhingran, 1991). Artificial breeding in captivity was conducted between 10.00 and 10.30 pm by injecting ovaprim, a synthetic inducing agent (combination of salmon gonadotropin-releasing hormone analog (D—Arg 6 Trp 7 Leu 8 Pro 9-Net) and domperidone dissolved in calibrated quantities of nontoxic organic solvent) (Nandeesha et al., 1990), a product of Syndel Laboratories, Canada, was administered to the brooders

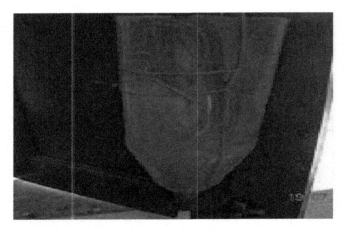

Figure 7.17 A single Thums Up bottle with fertilized egg.

intramuscularly in the caudal peduncle above the lateral line at a 45-degree angle. The required doses were 0.3 and 0.5 mL/kg body weight for male and female fish, respectively. Both male and female fish were given a single dose and the induced fish (male and female) were kept in separate aquaria. A sex ratio of two males to one female was maintained.

On the very next day, between 5 and 6 o'clock in the morning, eggs were stripped out into a dry and clean stainless steel plate and subsequently milt oozed out over the eggs. The milt and eggs were mixed thoroughly with the help of an avian feather. A little freshwater was added to facilitate the fertilization process. The fish indicate their readiness for stripping by rubbing their bodies against the wall (cloth) of the hapa (Figs 7.17–7.19).

7.19.1.7 Hatching Operation

Fertilized eggs were released into plastic Thums Up bottles (2 L capacity) at 500 ± 50 per unit for hatching. Bottles were arranged upside-down in a row over a wooden platform with holes depending upon the number of bottles while the bottom end was attached linearly to an iron rod. The mouth of each bottle remains inserted into a hole on a wooden plank while a slice is removed from the bottom end. The mouth at the lower end remains connected to a water flow by a rubber tube connected to the water line with a screw valve to control flow and to maintain constant circular upside-down motion within the bottle. This can be termed as a mini glass jar hatchery, easy to handle at the backyard with a lower requirement for water and ensuring a higher hatching rate (Fig. 7.13). This is a very low-cost

Figure 7.18 Demonstration of low-cost hatching unit to farmers.

Figure 7.19 Sets of inverted Thums Up bottles (miniature glass jar hatchery).

glass jar hatchery for poor farmers. On campus training at KVK, Murshidabad, farmers are well trained and shone in its operation. During the incubation period water quality parameters were tested following a standard method (APHA, 1995). The whole cycle of the hatching experiment was repeated thrice, in order to overcome methodological errors and the results obtained from all three sets of experiments were statistically analyzed.

In the present study, both male and female *C. reba* injected with ovaprim responded well to stripping with complete release of eggs. The rate of fertilization was 86%. Fertilized eggs were demarsal,

Table 7.9 Relative Fecundity of Female Raikhor Bata at the Dose of 0.3 mL/kg Body Weight

Observation No.	Average Fecundity (No. of Ova)
1	6780
2	4250
3	7181
4	9023
5	5042
6	3254
7	7374
8	8256
9	8691
10	9120
11	5680

nonadhesive, nonfloating, and transparent. Water-hardened eggs were spherical in shape and 2 mm in diameter. Twitching movements of developing embryos within the fertilized eggs started by 12 hours and eggs gradually transformed into elongated tubular larvae or spawn. Hatching took place 8–10 hours after fertilization at 27°C, with a success rate of 91.6%. Newly hatched spawns are transparent and without any pigmentation, which appears after 5–6 hours. At hatching body length measured 2.75 mm.

7.19.1.8 Physicochemical Parameters

Physicochemical parameters of water during incubation and hatching recorded a more or less congenial temperature of 27°C, dissolved oxygen 5.5 mg/L, and pH 7.6. Temperature: 26.5–27.8°C, pH: 7.2–7.4, dissolved oxygen: 5.2–6.4 mg/L.

7.19.1.9 Relative Fecundity (Number of Ova)

In this experiment, it was observed that the relative fecundity varied in different observations at a 0.3 mL/kg body weight dose (Table 7.9). Fecundity was highest at the 10th observation as 9120 (number of ova) and the lowest fecundity was observed at the sixth observation as 3254 (number of ova) (Fig. 7.20).

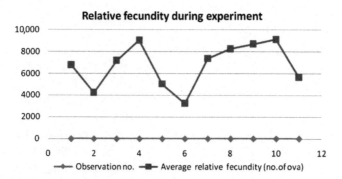

Figure 7.20 Relation between relative fecundity and different observations.

Table 7.10 Fecundity in Relation to Egg Diameter

Fecundity (No. of Ova)	Fertilized Egg Diameter (mm)
6780	2.6
4250	3.0
7181	2.6
9023	2.5
5042	2.8
3254	3.0
7374	2.7
8256	2.5
8691	2.5
9120	2.5
5680	2.8

7.19.1.10 Egg Diameter (mm)

Fully developed eggs (10–15) were collected in a 5-mL plastic test tube containing Simpson solution (2 mL) as preservative. The swollen egg diameter was measured by ocular micrometry (Table 7.10; Fig. 7.21).

7.19.1.11 Spawn

In this experiment, it was observed that spawn length (mm) was varied with relative fecundity (Table 7.11). The spawn length (mm) and relative fecundity were negatively correlated. It was shown that

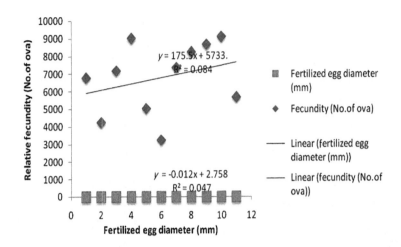

Figure 7.21 Relation between absolute fecundity (no. of ova) and egg diameter (mm).

Table 7.11 Length of Spawn as Observed During Experiment

Fecundity (No. of Ova)	Spawn Length (mm)
6780	3.2
4250	3.4
7181	3.3
9023	3.1
5042	3.0
3254	3.2
7374	3.2
8256	3.1
8691	3.2
9120	3.0
5680	3.2

there is an inverse relation between relative fecundity (number of ova) and spawn length (mm).

7.19.1.12 Description

In this present study, induced breeding was done using ovaprim, a synthetic hormone, at 0.3 mL/kg of body weight and 0.5 mL/kg of body weight for males and females, respectively.

7.19.1.13 Sustainability

Though the technology is cost-effective, eco-friendly, and very much effective in the seed production of minor carp, it bears lower number of eggs compared to major carp. Continuous trials along with determination and an innovative approach have led to the development of another modification of the glass jar hatchery by fish breeders, which is very effective in increasing the hatching rate to 80%. This method removes the constraints of quality seed production in captivity with little investment. The fish breeders, interestingly, add another modification to the glass jar as described below.

7.19.2 Further Modification of the Glass Jar Hatchery for Increased Hatching of Black Carp & *P. sutchi*

7.19.2.1 Initial Stage

The fish breeders of Bengal (an eastern state of India), realizing the uneconomic approaches of the breeding pool, started exploring other alternatives which would be economic and particularly would reduce water use. In this endeavor the fish breeders introduced a conical cement jar which is not only economic but easy to handle and increases production by increasing the fertilization and hatching rates. This cement jar hatchery was first introduced in 2000 for hatching of pangasid cat fish eggs as the fish breeders found difficulty in undertaking hatching of eggs in the hatching pool.

The skeleton of the conical jar is made up of iron installed in a frame. The length of the jar measures 34 inches except the bottom conical part, which is 22 inches long while sides are 24 inches. A 1-inch diameter inlet pipe is attached to the rear conical part of the jar and remains connected by a 3-inch diameter pipe ultimately connected to an overhead tank. An outlet pipe (conduit) of 3-5 inches diameter is attached to a jar just below the upper circular margin. A fine-meshed circular net of 22 inches length remains attached to the jar on its inner side and the other end (upper) fastened to a circular iron frame. The net maintains an erect posture above the jar as the iron frame remains tied with strong thread to a pipe overhead (Figs 7.22-7.27).

Figure 7.22 Arrangement of fine-meshed net within cement jar to outflow of excess water.

Figure 7.23 Series of jars with fine-meshed net above the jar tied to a rod above.

7.20 New Dimension in Common Carp Breeding

7.20.1 Nature of Common Carp Eggs

Common carp eggs possess an adhesive layer covering the shell, composed of glycoprotein (a compound of sugar and protein) and

Figure 7.24 Water inlet to jar.

Figure 7.25 Screw valve to adjust flow of water within the jar.

Figure 7.26 Construction of jar at the initial stage.

Figure 7.27 An inverted iron jar at the initial stage.

become activated when the egg comes into contact with water. The adhesive layer sticks the eggs to some objects or to each other. Depending on the nature of attachment the eggs are of two types:

1. Eggs sticking to objects
2. Eggs sticking only to each other, forming a clump or egg mass.

Again, the degree of adhesiveness varies, sometimes so strong that attached eggs suffer damage if separated from their substratum.

Sometimes the adhesiveness is quite weak and the eggs can easily be separated from the substratum, there are several grades or variations between these two extremes. However, the adhesiveness loosens gradually as the fertilized egg proceeds toward further development. Fertilized eggs also vary in size and the factors which control the size are egg "kernel," egg shell thickness, and the extent of perivitelline space. While the first two factors control the size of the "dry" eggs (i.e., before they come in contact with water), the third controls the size of the water-hardened eggs. Salinity exerts a negative influence on adhesiveness.

The sticky layer facilitates adhering of ovulated ripe eggs to submerged objects in water where they can develop. Unripe eggs do not possess such a layer. "Dry eggs" are also not sticky, but become sticky in contact with water. Adhesiveness is strongest at the initial stage and becomes dissoluble with the progress of time. Stickiness varies with the species, it is very weak in pike, fairly strong in cyprinids and pike–perch, and strong in the European catfish. This may be due to varying compositions of the adhesive glue in different groups. It is known now that salt water and common salt solution deactivate the adhesive glue of the carp egg, while carbamide and guanidin dissolve the sticky material from the egg surface. As the eggs ripen in contact with fresh water, they form clumplike balls, through the activation of adhesive glue or possibly its remnants on the egg surface. Therefore, it is recommended to wash the eggs thoroughly with a salt–carbamide solution during the swelling of the eggs, in order to remove the sticky material from the surface. A still effective and easier technique consists of quick washing with a weak tannin solution at the end of the swelling as we know that tannin denatures all protein compounds and thus helps eliminate the adhesive glue. For removing adhesive glue the eggs are first washed repeatedly during their swelling stage with increasing quantities of the salt–carbamide solution (*fertilizing solution*) and then washed three or four times in a tannin solution.

Fertilization of ripe egg is limited as the micropyle is closed as soon as the egg comes in contact with water because of immediate swelling of the egg following fertilization. Both in the cases of common carp and Chinese carp, closure takes place within only 45–60 seconds, providing very little time (maybe a few seconds) for the sperm to penetrate the egg.

7.20.2 Water Quantity: An Important Criteria for Successful Fertilization

During artificial fertilization and stripping, the amount of water or fertilizing solution plays a crucial role in the fertilization rate. When the amount is more the sperm are likely to stray away from the micropyle. On the contrary, if the water is insufficient, the micropyle of one egg is covered by another as the eggs stick together to form a ball-like structure, resulting in fertilization of the eggs at the surface of ball of eggs despite the presence of innumerable sperm in the stripped milt; about 10–20 billion in one cubic centimeter of milt, meaning the sperm fail to fertilize 80% of the eggs. As the eggs become clumped so swelling and subsequent development of eggs are impaired and total mass is destroyed due to anoxia. It has been seen that the virility of sperm lasts much longer in carbamide–salt solution (20–25 minutes) than in water (1–2 minutes).

Fertilizing solution is prepared by dissolving 30 g carbamide (urea) and 40 g common salt (NaCl) in 10 L of clean (preferably filtered) pond water. Fertilization solution at the rate of 10–20% of the volume of the eggs is added and the mixture is stirred with a feather for about 3–5 minutes continuously, for thorough mixing of sperm and eggs. Subsequent stirring is carried out by hand. One person can stir two bowls continuously, for 4–6 minutes by rotation.

The eggs absorb the solution without any harm and begin to swell. Repeated washing is needed to remove the adhesive glue completely. Common carp eggs swell to about 10 times their original size during the process of water hardening, for 0.2 L of dry eggs a 3-L bowl is needed to provide sufficient space for optimum swelling. The eggs are then treated with a second solution, prepared by mixing 5–8 g of tannin in 10 L of water. Solution should be freshly prepared every time it is required.

About 2–4 L of tannin solution is placed in a plastic bucket and a maximum 2–3 L of swollen eggs are placed in the tannin solution with continuous stirring for 3–5 minutes. The eggs are then washed thoroughly with clean water. As the eggs settle down excess water is drained out. After stirring for 3–5 seconds, clean water is added. The eggs are treated again with 1–2 L of tannin solution with subsequent washing. As tannin is toxic the egg should be washed with clean water repeatedly. Again when tannin solution is inadequate, the eggs again exhibit adhesiveness. At this stage water-hardened eggs can be separated from each other by hand. The said procedure is applicable for treating sticky eggs of tench (*Tinca vulgaris*), aspius (*Aspius aspius*), bream (*Abramis brama*), and other cyprinids, to remove the adhesive glue from the surface of the egg.

7.20.3 Fertilization of Nonsticky Eggs

The absence of a sticky layer renders the fertilization and handling of nonsticky eggs a much easier operation. Clean water only needs to be used and there is no necessity for any "fertilizing solution." Here also the water to be added to facilitate fertilization is about 10–20%

of the volume of the "dry" eggs. The process lasts for about 5 minutes, during which time stirring of the eggs should be continuous. The eggs are next transferred to incubators, taking care to put in only as many eggs as can be accommodated in the container, keeping in mind that they swell about 40–60 times their original size (Fig. 7.12). Some fish culturists prefer to wait until the eggs are fully swollen in the bucket itself before they are transferred to the incubators. However, this operation may result in damage to a number of eggs. Therefore, it is better and more convenient to transfer the nonsticky eggs to the incubators immediately after fertilization and before they begin swelling.

7.20.4 The Role of Carbamide Solution as a Catalyst in Fertilization

It is established now that carbamide solution increases capacitation as well as the viability of sperm 10–20 times longer than normal freshwater. In addition, the solution increases the fertilization rate and helps dissolve the adhesive materials which clog the micropyle of eggs. In the case of pike (*Esox lucius*), carbamide solution (16 g carbamide dissolve in 1 L water) increases the fertilization rate very significantly, up to as much as 80–90%. However, the optimal concentration differs from species to species. An optimal concentration for a particular species is that which provides maximum vigor to sperm for a period of 10–15 minutes. It indicates that carbamide fertilizing solution acts as a catalytic agent in increasing fertilization, particularly during stripping. Common salt solutions exhibit similar properties during artificial fertilization of pike.

7.20.5 Swelling of Eggs

It has been observed that not only ripe eggs but preovulatory and postovulatory eggs are found to swell on contact with water or "fertilizing solution" though the degree of swelling varies. This indicates that for water hardening fertilization is not needed. As the swelling starts, the micropyle closes, preventing entry of sperm. Both in the cases of common carp and Chinese carp eggs, this happens within 1 minute of spawning and their contact with water. This indicates that the ripe eggs, after spawning, should be fertilized within 1 minute and considering the same the stripped eggs should be fertilized with stripped milt within 1 minute.

The water-hardened eggs consist of (1) a central germ disk, (2) perivitelline space, and (3) the egg shell. The germ disk consists of a presumptive cell mass for the embryo proper, such as also yolk mass, and fats. Two poles are distinct, namely, the upper animal pole or blastodisc, and the ventral vegetative pole or yolk mass. The animal pole includes the cell nucleus, which has by then the normal number of chromosomes as in any somatic cell. Around the kernel is the so-called perivitelline space, which is filled with perivitelline liquid. This liquid contains dissolved proteins. The egg is enclosed in an egg shell, which consists of one, two, or three layers in different fish species. The thickness, hardness, and other characteristics of the egg shell may also vary with different species. The type of incubator, therefore, will have to be selected depending on the nature of the egg shell. The egg shell is very delicate in some fish and ruptures easily, leading to the spoilage of the egg. In other fishes, it is very tough and can hardly be ruptured even by pressing between the fingers (e.g., mahaseer).

7.20.6 Selective Breeding of Common Carp

Qualitative genetic traits studied in common carp included the inheritance of scale pattern ("scaly" SSnn, Ssnn; "mirror" ssnn; "line" SSNn, SsNn, and "leather" ssNn) and the lethal/deleterious effects of the N allele (Kirpichnikov, 1981), types of pigmentation (wild-type, black, gray, blue, gold, orange, red; David et al., 2001) used as genetic markers or for selective breeding of colored breeds, and pleiotropic effects of genes responsible either for scale patterns or for coloration on various biological and productive characteristics. Compared to fully scaled and/or wild-type colored carp, those with other scale and/or color patterns mostly exhibit reduced growth, survival, and disease resistance. Quantitative traits studied involved growth rate, disease resistance and cold resistance (Vandeputte, 2003; Hulata, 1995) with mostly low to intermediate heritability (h^2) estimates and burdened with environmental biases (Vandeputte, 2003). Selective breeding for disease and cold resistance resulted in developing several breeds (e.g., krasnodar carp, Kirpichnikov et al., 1999; ropsha carp, Kirpichnikov, 1999), while simple mass selective breeding for growth did not show improvement in the line selected for faster growth (Hulata, 1995). Population genetic studies with allozymes and/or microsatellites revealed lower variability of domesticated breeds compared to wild populations (Kirpichnikov et al., 1993; Kohlmann and Kersten, 1999; Kohlmann et al., 2005) and low genetic distance between breeds (Kirpichnikov et al., 1993). It indicated that many breeds have been established using a small effective number of broodstock, which has resulted in some inbreeding and which might hamper possibilities of genetic gain from selective breeding (Kirpichnikov et al., 1993). With construction of synthetic strains to start a within-strain selective breeding program with a sufficient number of families and standardized family size by separate rearing of families until fry mortality stops, or with parentage assignment by means of microsatellites, more efficient breeding programs may be designed (Kohlmann et al., 2003; Kohlmann and Kersten, 1999; Vandeputte et al., 2004). Common carp have been subjected to all kinds of chromosomal manipulations (Gomelsky, 2003). Gynogenesis, both meiotic and mitotic, revealed increased homozygosity (with inbreeding coefficients $F = 0.6$ and 1.0, respectively) and female homogamety XX. Mitotic gynogenesis was used to produce clones. Androgenetic YY males were crossed with normal females to produce all-male progenies. Gynogenetic progeny subjected to hormonal sex reversal resulted in production of XX neo-males, and crossing these with normal females produced all-female progenies. Rearing the female monosex stock enhanced production by 7–8% (females being 15% heavier than males), in tropical/subtropical conditions when fish reached sexual maturity before market size. However, in a European temperate climate the female monosex stock grew better and had better slaughtering value only in the first 3 years, but not at market size. Triploids are characterized by a reduction in gonad development but not with increased somatic growth. Microsatellite DNA markers were developed and applied in studies of genetic variation and diversity (Desvignes et al., 2001), parentage assignment (Vandeputte et al., 2004), and a genetic linkage map was constructed (Sun and Liang, 2004). Growth hormone (GH) gene transfer was described and the technique was developed to enhance common carp production in China (Wu et al., 1993), firstly using human GH and later using grass carp GH fused to common carp β-actin promoter. The transgenic fish showed higher growth performance and food conversion efficiency than the controls, but no transgenic fish have been commercially approved for human consumption (Hulata, 1995; Fu et al., 2005). Sterile triploid transgenic fish were produced to avoid environmental impacts. Most of the world production is carried out using unselected strains (Vandeputte, 2003). When they exist, the breeding program is mostly based on crossbreeding as it brings quick improvement of growth performance (heterosis effect) in F_1 generation. It is widely used in Hungary, Israel, Czech Republic, and other countries. Crossbreeding of breeds developed from both subspecies (*C. c. carpio* and *C. c. haematopterus*) improved the survival rate of fry, and disease and cold resistance. However, improper use of hybrids for further breeding brought contamination to the purebred stocks (Flajšhans et al., 1999). Live gene banks of common carp breeds are kept and new forms are continually tested, e.g., in the Research Institute for Fisheries, Aquaculture and Irrigation, Szarvas,

7.20.7 Characteristic Soil Discovered Bengal Breeders, India, Effective in Removing Adhesive Glue and in Increasing Fertilization and Hatching

One type of characteristic soil has been discovered by the author from certain districts of West Bengal, India, which bears properties to remove adhesive glue from the adhesive eggs. One handful of soil is enough to dissolve adhesive glue from the eggs of two stripped female, when placed in 10 Liters of water in a hundi. This procedure is not only economic but leaves no toxic effect on the egg as happened with tannin solution.

7.20.7.1 Brood Stock Collection and Management

Collection of brood stock in adequate number, cataloging of their geographical origin, their genetic characterization, and maintaining their pedigree record are important prerequisites for the breeding program. These aspects are of great genetic relevance for the success of any breeding program. Further proper feeding of brood stock and their health maintenance are important management aspects.

7.20.7.2 Artificial Breeding

Artificial breeding has hormonal and genetic components. Treatment of appropriate dosage of gonadotropin (crude pituitary extract) or any other inducing agent like gonadotropin-releasing hormone (GnRH) analog (ovaprim) is required to obtain gametes for artificial fertilization. Artificial fertilization can be done by hand-stripping or seminatural breeding. Hand-stripping involves mixing of milt and eggs on a Petri dish or tray for artificial fertilization. This process is laborious. On the other hand, seminatural breeding allows the hormone-induced fish to spawn on their own in a "breeding pool." This type of spawning practice is more convenient and economical.

To find out a cheap and specific chemical we have already conducted some study at field. We have discovered a specific soil from a certain region of Bengal, India, which is very effective in removing the adhesive component. When a handful of soil is mixed with 10 L of water in a handi and the eggs produced out of two successive games are added to it, individual eggs separate from each other.

7.20.7.3 Expected Outcome

Although common carp is an important item of stocking for composite farming, often farmers do not consider the species for farming due to nonavailability of seed from any source. Actually the fish seed producers of India, particularly of Bengal, due to nonavailability of seeds from natural sources take the risk of common carp seed production through hypophysation or stripping. This is because of the lack of any cost-effective technology for degumming and thereby enhanced fertilization. The fish farmer as of now depends mainly on wild seed for stocking. In this context the proposed study will open a new dimension for standardized seed production of common carp in the farming sector and at the same time this will increase production

Figure 7.28 *Cyprinus* eggs laid on the roots of submerged weeds.

through composite farming due to easy availability of seed. In this context it may be mentioned that Bengal farmers are now pioneers in seed production of *Pangasius* in India (Fig 7.28).

Common carp are a multiple breeder that produce adhesive eggs. After sex play, the female lays eggs on submerged weeds or other suitable substrate and the adhesive glue help eggs attach to the substratum where they are then fertilized by the milt released by the male. On contact with water, the eggs immediately assume the shape of round balls before being fertilized. With the passing of time, agglutination increases further. This also happens with *P. sutchi*, a handsome ornamental indigenous catfish that lays adhesive eggs. This transformation of eggs is due to the interaction of the adhesive or gumming substances with water. In both normal breeding and hatcheries, the sperm fail to penetrate the adhesive covering surrounding the eggs and fertilize the eggs only at the periphery of the egg balls. This results in decreased fertilization and poor availability of hatchlings.

7.20.7.4 Degumming Eggs

To increase fertilization rates, several procedures have been developed for degumming of eggs. Some authors advocate the use of tannin in a solution of 15 g tannin/10 L water. However, tannin is poisonous to fish eggs, and its prolonged use can generate various negative impacts. An alternative method is the use of a urea–sodium chloride solution, but this approach is also ineffective and there is potential for contamination. The use of amul milk at a solution of 200 g/10 L water is effective, but is a costly approach.

7.20.7.5 New Procedure

In field studies to find an inexpensive, yet effective treatment to remove adhesive glue from eggs, the author discovered that some soils from a certain region of Bengal, India, proved effective in removing the adhesive component. Eggs and milt were collected through stripping followed by fertilization by use of a feather. The fertilized eggs were placed in a tray containing 10 L of water into which a handful of the soil is mixed. The water/soil solution remove the adhesive glue from eggs produced by two successive games (spawning). Individual eggs separated from each other, and fertilization was enhanced. After thorough mixing and separation of individual eggs, the eggs are stocked into the tank for further development. The released mixture adds a yellowish tint to the water.

7.20.7.6 Perspectives

Although common carp is an important species for stocking in multispecies operations, farmers generally do not consider these species for more intensive farming due to the lack of availability of seed. Fish farmers depend mainly on wild seed for stocking carp. Some seed producers in India, particularly in Bengal, also take the risk of common carp seed production through stripping. The soil-based approach for removing carp egg adhesive to improve fertilization opens a new dimension for standardized seed production that can help increase production. This technology is also being studied with, for which increased fertilization and hatching rates are expected.

7.21 Captive Breeding of Two Recently Introduced Carnivorous Fish Species in West Bengal

7.21.1 Induced Breeding of PACU (*P. brachypomus*) in Captivity With Pituitary Extract

7.21.1.1 Introduction

Pacu (*P. brachypomus*), a native fish of the Amazon river of South America was recently introduced into India through Bangladesh and has added one more species to the list of alien introductions (Fig. 7.29). A tendency has been observed among the fish breeders of Bengal to raise seeds from alien species, mostly carnivorous in nature. Besides breeding individual species, fish breeders are also interested in conducting hybridization programs between Indian major carp and exotic species, and are unaware of the genetic consequences of their

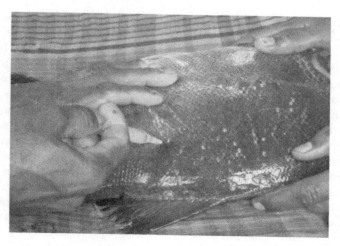

Figure 7.29 Administration of pituitary dose.

breeding programs. These fish were introduced in India probably in 2003–2004 and leading hatchery owners started culturing in captivity to raise brood fish for breeding programs. The fish attain maturity at 3+ stage with a stocking density of 2000–2500/bigha. The fish are stocked at the ratio of 1 pacu:3 IMC when polyculture is undertaken.

Some farmers reported that polyculture gives better results than monoculture. During brood stock management the fish are fed with oil cake, boiled rice and grains, peas, and fishmeal at 5% of the total biomass. Compared to other food ingredients fishmeal and groundnut oil cake offer better growth and maturity. The maturity of males and females coincides during the breeding season. One to two days before breeding, feeding is stopped. The mature male and female fish are induced with pituitary extract. The first dose for females is 2 mg/kg body weight and after 4.5–5 hours the second or resolving dose is administered at 10–12 mg/kg body weight (Fig. 7.38). During the second injection to females, males are given a single resolving dose of 2 mg/kg body weight. After injecting the fish are kept in hapa, and after 5–6 hours stripping is done. After injection, the injected male and female fish are kept in a cemented cistern. The induced fish produce a characteristic vibrating sound on the walls of the cistern. The eggs are stripped out in trays (Fig. 7.16) and milt is mixed with eggs using a feather and by addition of water. The ratio of males to females for stripping ranges from 1:1 or 1.5:1, depending on the maturity and density of milt (Fig. 7.35).

Environmental factors play an important role in both the dose requirement and stripping of fish. When the temperature is higher than 34–35°C, then both the dose requirement and stripping time are less. Again, for better fertilization a congenial environment is needed. The fertilized eggs are transferred successfully into two hatching jars in 48 hours and in between the hatchlings are kept in hapa. As the yolk is absorbed the hatchlings become longer. From the third day onward the hatchlings are ready for sale. For nursery management the spawns or hatchlings are stocked in the nursery pond, prepared beforehand, at the rate of 1.5 lakhs/bigha. The spawns are fed with artificially prepared feed out of groundnut oil cake, rice polish, and soya bean dust, and also on zooplankton. Generally 2 kg feed/bigha every day is given with the above stocking rate. Within 8–10 days the spawns developed into fry, which are transferred to a rearing pond. A three-bigha pond is ideal for rearing, and the stocking density is 1 lakh/bigha. The fry are fed with artificial food at the rate of 5 kg/bigha every day, made out of groundnut oil cake, sunflower oil cake, and fish meal, and also with boiled rice and Canadian peas.

7.21.1.2 Future Prospects

These fish were introduced into the farming sector of Bengal not more than 5 years before the time of writing. At this stage two or three farmers have already standardized the captive breeding and culture of this fish. At the fingerling stage the fish possess an alluring shining color with a tint of blood red color around the ventro-anterior region of the trunk. At maturity the color becomes dull with some round spots appearing all over the body. The overall color is dark-ash in nature with a shape like pomfret. Within a short period there has been a great demand for spawn and fry from Midnapore and 24 Parganas (south) districts of the state. Demand for seeds is greater in Chennai, Kerala, and Andhra Pradesh. The fish seed cultivators of the said regions supply the mature fish to Bihar, Uttar Pradesh, Bangladesh, and Pakistan, where people are very fond of this fish as it possesses more meat and fewer spines. The characteristic pomfret-like shape is considered as one quality customers prefer. These fish have already established a place in the farming sector of Bengal and a good number of freshwater hatcheries will be converted into pacu hatcheries in the near future (Figs 7.29 and 7.30).

7.21.2 Induced Spawning of *P. sutchi* With Pituitary Extract

The main species of pangasid catfish recently adopted for culture with Indian major carp are yellowtail catfish (*Pangasius pangasius*) and sutchi catfish (*Pangasius sutchi*) (Fig. 7.31). These fish were introduced into the farming system of Bengal from Thailand through Bangladesh in 1994–95. Though carnivorous at an early stage, the fish are compatible with Indian major carp from 5 days onward and can

Figure 7.30 Injection of male pacu brooder.

Figure 7.31 Pangasius catfish. This specimen comes from a farm in Myanmar, close to Yangon.

grow to 3 kg/year on a balanced diet (Rahaman et al., 1991, 1992). These fish have already established their importance as profitable species in the aqua farming of Bengal. As a result of its remarkable growth rate (almost 1 kg in 90 days), there is now much enthusiasm among the fish breeders and farmers of Bengal for its artificial spawning and culture. The demand for its seed is increasing rapidly. In view of the increasing demand for *P. sutchi* seed we tested techniques for induced spawning and larval rearing of this fish (Fig. 7.32).

Figure 7.32 Mature female *Pangasius sutchi* taken for injection.

7.21.2.1 Technique for Induced Spawning

Brood fish were raised in farm ponds (area 2500 m³) from fry stage using a high-protein balanced diet composed of cereal waste (25%), rice bran (20%), mustard oil cake (15–20%), broken grain (25%), and animal meat (10–15%). The diet was provided two or three times per day at the rate of 5% of body weight. To check the growth rate the percentage of animal meat was reduced as per requirement. The fish attain sexual maturity at 4 years when they normally reach a size of 7 kg. However, for the convenience of breeding the weight of brood fish we used was restricted to 1.5–2.0 kg with intensive stocking. Males and females are easily distinguished, particularly around April. Egg-bearing females are identified by their big, soft and distended belly with swollen and reddish pink vent (Fig. 7.3). Males can easily be identified by their reddish genital opening, and oozing of milt when the abdomen is pressed. As with clarid catfish only carp pituitary extract (CPE) was used as the agent for inducing spawning. The results were promising. There are also reports regarding the successful use of human chorionic gonadotrophin (HCG) and LRH-A in combination (Rahaman et al., 1993; Saidin and Othman, 1986). A stimulatory first dose of 1.5 mg CPE/kg body weight was injected into mature females (Fig. 7.33). After 5–6 hours the second resolving dose of 6 mg CPE/kg body weight was administered to females. Males were injected at the rate of 1 mg CPE/kg body weight at the same time as during the second injection to females. In the case that female broodstock failed to reach the peak of maturity, the stimulatory dose would

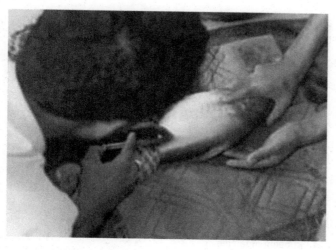

Figure 7.33 First injection to the female.

be increased to 2.2 mg CPE/kg body weight. The resolving dose in such situations would be 9–10 mg CPE/kg body weight. Males were given a single resolving dose of 2 mg/kg body weight at the time of second injection to females. Variations in environmental temperature have a strong effect on the effectiveness of the dose. When the temperature rises above 30°C less CPE is required and more is needed when the temperature falls below 28°C. Breeding starts from April and continues until mid-September. One brooder can be used at least two times during the same breeding season. After injection the fish were returned to their respective cement tanks or hapa. Spawning occurs after an interval of 5–6 hours. Both natural spawning and stripping are possible, but as the eggs are adhesive in nature stripping was considered (Figs 7.34 and 7.35).

We rinsed eggs in milk powder solution in aluminum hundi to remove the adhesive gelatinous covering of the fertilized eggs (Figs 7.36 and 7.37). We prepared the milk solution by adding 200 mL of milk in 30 L of water for 20 minutes. Afterwards the fertilized eggs were transferred to a Chinese hatchery (Table 7.12).

7.21.2.2 Effectiveness of the Technique

In all trials, the fish responded positively and ovulated within 5–6 hours after the second injection. The fertilization rate ranged from 95% to 100%. The fertilized egg does not swell as it does in carp and hatched within 24 hours at temperature ranges between 30°C and 32°C. Temperature was a prime factor for fertilization and

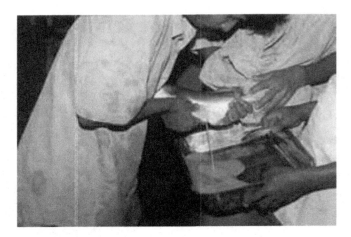

Figure 7.34 Stripping of female.

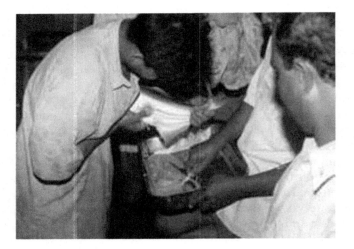

Figure 7.35 Stripping of male.

hatching. There are several other reports of the successful breeding of *P. sutchi* in Indonesia and Thailand. According to Saidin and Othman (1986) the hatching period ranged between 24 and 26 hours at a water temperature of 28–32°C with ovulation occurring in between 70% and 80% and with survival of hatchings from 30% to 45%. Milt from one male is sufficient to fertilize the eggs of three to four females. The dry method of egg fertilization was followed. It was observed that the hatchlings became cannibalistic if sufficient food was not available after 3 days of hatching. We fed our hatchlings on

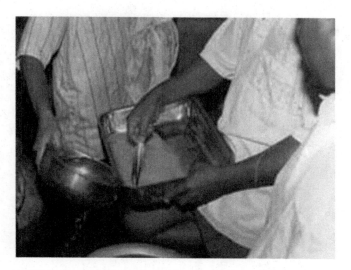

Figure 7.36 Eggs and milt of mixed carnivorous species in West Bengal.

Figure 7.37 Captive breeding of two recently introduced carnivorous fish.

lactogen for the first 48 hours. The hatchlings become carnivorous from about 72 hours and at this stage weighed 500 mg. We fed earthworm dust three times a day continuing up to 5–8 days. After 10 days we fed soyabean dust as supplementary feed. Afterwards transferred to a rearing pond with natural feed.

Table 7.12 Positive Fish Response to Trials After Second Injection

Date	No. of Females per Trial	Average Weight of Females (kg)	Dose of Pituitary Extract per kg Body Weight				Response to Treatment (Ovulation + Fertilization) (%)	Hatching (%)
			Male	Female				
				1st	2nd			
5/4/2000	3	1.5	2.0	1.5	8.5		70–89	90–92
9/4/2000	3	2.0	2.0	1.5	8.5		75–90	90–92
15/4/2000	2	2.5	2.0	2.5	8.0		80–96	95
20/4/2000	3	1.8	2.0	1.5	8.5		80–89	95
25/4/2000	2	1.7	2.0	1.5	9.0		85–98	96
28/4/2000	3	2.5	2.0	2.5	8.0		85–92	90

7.22 Innovative Technology Developed by Fish Breeders

7.22.1 Addition of Fertilization Pool and Rejection of Breeding Pool

As we know the technology of seed production in captivity was adopted by Bengal farmers on their own and during this period they have developed several trial and error methods for further refinement and/or modification of the technology in vogue in the state. The establishment of an innumerable number of hatcheries and successive competition led them to try out new ideas which are easier and more economic at the same time. To reduce the water budget in the breeding pool they rejected the breeding pool concept and at the same time developed the concept of a fertilization pool (egg collection jar) with an even-shaped floor, in between the breeding and hatching pools. This technology was not accepted by other fish breeders as it failed to show economic viability.

7.22.2 Single Row of Duck Beak Inlets in Hatching Pool

In some hatcheries, fish breeders installed a single row of duck beak inlets at the floor of the incubation pool, but failed to make it viable and was of little interest among fish breeders (Fig. 7.38).

7.22.3 Installation of Hatching Pool in Pond to Compromise With Water Budget

As the breeding operation involves a huge requirement for water, so fish breeders, from the very dissemination of the technology in

Figure 7.38 Single row of duck beak inlet.

Figure 7.39 Chinese hatchery constructed in pond to reduce water budget.

Figure 7.40 Fertilization pool.

the field, were tirelessly involved in developing an alternative technology which would reduce the water budget. Some hatchery owners installed hatching pools in ponds (Fig. 7.39) but this was not accepted by other hatchery owners (Fig. 7.40).

7.22.4 Low-Cost Technology for Construction of an Overhead Tank

The fish breeders and hatchery owners of Bengal added a new economic approach for construction of overhead tanks by curtailing the construction cost. In this endeavor they, while excavating a new pond, construct a mound out of the soil removed during excavation. The overhead tank is then constructed over the mound, which curtails the costs for construction of six to eight pillars (Fig. 7.41).

Figure 7.41 Low-cost technology for construction of overhead tank.

7.22.5 Technology for Intermittant Removal of Undesirable and Decayable Matter of Hatched-Out Eggs Without Disturbing the Spawn

The fish breeders of Bengal, in their attempt to remove decaying matter of the hatched-out eggs from the incubation pools, place a stick on the surface in between the two chambers. Twelve to thirteen piece thick thread, made out of coconut fiber, remains hanging straight into the water as they carry weight at the lower end (Fig. 7.42). The newly hatched spawn move through the thread undisturbed, but other undesirable and decayable matter, due its sticky nature, attaches to the thread. After a certain interval the stick and the adjacent thread are taken out and after thorough washing are placed again in the incubation chamber. This procedure is repeated until the end of the hatching pool operation.

7.22.6 Enhancement of Water Hardening of Egg

The fish breeders of Bengal, to prevent premature hatching, treat the fertilized eggs with a special solution. The solution is added before breeding by placing in water for a month. The extract turns the water yellowish and is very effective in preventing premature hatching (Fig. 7.43). The solution puts a yellow coating (Fig. 7.44) over the egg and thus prevents premature hatching. This technology was later adopted by most of the fish breeders in Bihar state.

Figure 7.42 Innovative technology for removing sticky unwanted material of the hatched-out eggs.

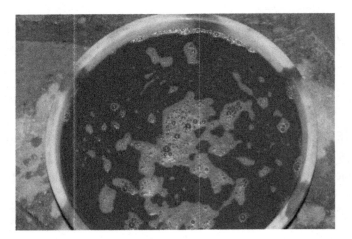

Figure 7.43 Solutions prepared from haritaki are used to prevent premature hatching.

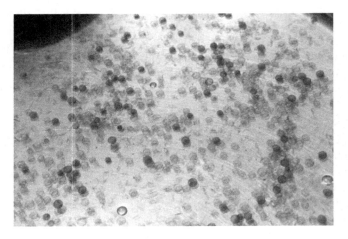

Figure 7.44 Fertilized eggs after absorbing haritaki extract.

References

APHA, 1995. Standard Methods for the Examination of Waterand Wastewater, 19th ed. American Public Health Association, Washington, DC.

Bhuiyan, A.L., 1964a. Fishes of Dacca, Asiat. Soc. Pakistan, Pub. 1, No. 13, Dacca, 21–22.

Bhuiyan A.L., 1964b. Fishes of Dacca, 1st ed. Asiat. Soc. Pakistan, 29–30.

Chatterjee, S., Isaia, M., Venturino, E., 2009. Spiders as biological controllers in the agroecosystem. J. Theor. Biol. 258, 352–362.

Chondar, S.L., 1999. Biology of Finfish and Shellfish. SCSC Publishers, pp. 66–75.

David, L., Rajasekaran, P., Fang, J., Hillel, J., Lavi, U., 2001. Polymorphism in ornamental and common carp strains (Cyprinus carpio L.) as revealed by AFLP analysis and new set of microsatellite markers. Mol. Genet. Genomics 266, 353–362.

Desvignes, J.F., Laroche, J., Durand, J.D., Bouvet, Y., 2001. Genetic variability in reared stocks of common carp (Cyprinus carpio L.) based on allozymes and microsatellites. Aquaculture 194, 291–301.

Flajshans, M., Linhart, O., Slechtová, V., Slechta, V., 1999. Genetic resources of commercially important fish species in the Czech Republic. Present state and future strategy. Aquaculture 173, 471–483.

Fleming, I.A., Kinder, K., Mjolnerod, I.B., Jonsson, B., Balsta, T., Lamberg, A., 2000. Lifetimes success and interactions of farm salmon invading a native population. Proc. R. Soc. Lond. 267, 1517–1523.

Fu, G., Vary, P.S., Lin, C.T., 2005. Anatase TiO2 nanocomposites for antimicrobial coating. J. Phys. Chem. B 109, 8889–8898.

Gomelsky, B., 2003. Chromosome set manipulation and sex control in common carp: a review. Aquatic Living Resources 16, 408–415.

Gupta, S., 1975. Some observations on the biology of *Cirrhinus reba* (Cuvier). J. Fish Biol. 7 (1), 71–76.

Gupta, S.D., Rath, S.C., Ayyappan, S., 2000. Designing and management of eco-hatchery complex for carp seed production. Fish. Chimes 19 (10&11), 27–38.

Hamilton, F., 1822. *An account of the fishes found in the river Ganges and its branches, Edinburgh & London*, 1–405.

Hansen, P.K., Ervik, A., Schaanning, M., Johannessen, P., Aure, J., et al., 2001. Regulating the local environmental impact of intensive, marine fish farming. II. The monitoring programme of the MOM system (Modelling-Ongrowing fish farms—Monitoring). Aquaculture 194 (1–2), 75–92.

Hulata, G., 1995. A review of genetic improvement of the common carp (Cyprinus carpio L.) and other cyprinids by crossbreeding, hybridization and selection. Aquaculture 129, 143–155.

Job, T.J., 1944. Madras Rural Pisciculture Scheme. Annual Progress Report to the I.C.A.R., Govt. Press, Madras.

Kallio-Nyberg, I., Koljonen, M.-L., 1997. The genetic consequence of hatchery-rearingon life-history traits of the Atlantic salmon (*Salmo salar* L.): a comparative analysis of sea-ranched salmon with wild and reared parents. Aquaculture 153, 207–224. doi:10.1016/S0044-8486(97)00023-9.

Kirpichnikov, V.S., 1981. Genetic basis of fish selection. Springer Verlag, Berlin, Germany, 1–410.

Kirpichnikov, V.S., Ilyasov, J.I., Shart, L.A., Vikhman, A.A., Ganchenko, M.V., Ostashevsky, A.L., et al., 1993. Selection of Krasnodar common carp (Cyprinus carpio L.) for resistance to dropsy: principal results and prospects. Aquaculture 111, 7–20.

Kirpichnikov, V.S., 1999. In: Billard, R., Reperant, J., Rio, J.P., Ward, R. (Eds.), Genetics and breeding of common carp. INRA, Paris, France, pp. 1–97.

Kohlmann, K., Kersten, P., 1999. Genetic variability of German and foreign common carp (Cyprinus carpio L.) populations. Aquaculture 173, 435–445.

Kohlmann, K., Gross, R., Murakaeva, A., Kersten, P., 2003. Genetic variability and structure of common carp (Cyprinus carpio) populations throughout the distribution range inferred from allozyme, microsatellite and mitochondrial DNA markers. Aquat. Liv. Res. 16, 421.

Kohlmann, K., Kersten, P., Flajshans, M., 2005. Microsatellite-based genetic variability and differentiation of domesticated, wild and feral common carp (Cyprinus carpio L.) populations. Aquaculture 247, 253–266.

McGinnity, P., Prodöhl, P., Ferguson, A., Hynes, R., Maoiléidigh, N.Ó., Baker, N., et al., 2003. Fitness reduction and potential extinction of wild populations of Atlantic salmon, *Salmo salar*, as a result of interactions with escaped farm salmon. Proc. R. Soc. London, Ser B 270, 2443–2450.

Menon, A.G.K., 1999. Check list—fresh water fishes of India. Records of the Zoological Survey of India, Occasional Paper No. 175.

Mookherjee, H.K., Sen Gupta, S.N., Roy Choudhury, D.N., 1946. Food and its percentage composition of the common adult food fishes of Bengal. Sci. Cult. Calcutta 12 (7), 247.

Nandeesha, M.C., Keshavanath, P., Varghese, T.J., Shetty, H.P.C., Rao, K.G., 1990. Alternate inducing agents for carp breeding: progress in research. In: Keshavanath, P., Radhakrishnan, K.V. (Eds.), Proceedings of the Workshop on Carp Seed Production Technology, September 2-4, 1988. Special Publication 2. Asian Fisheries Society, Indian Branch, Mangalore, India., pp. 12–16.

Rahman, A.K., 1989. Freshwater fishes of Bangladesh. The Zoological Society of Bangladesh, Bangladesh, p. 352.

Rahaman, M.K., Mazid, M.A., Rahman, M.A., Akhter, J.N., 1991. Formulation of quality fish feeds from indigenous raw materials and its effect on the growth of Catfish *Pangasius pangasius* (Ham.). J. Zool. 6, 41–48.

Rahaman, M.K., Akhter, J.N., Mazid, M.A., Halder, C.G., 1992. Culture feasibility of exoitic catfish *Pangasius sutchi* (Fowler) in freshwater ponds of Bangladesh. J. Inland Fish. Soc. India 25 (2), 26–30.

Rahaman, M.K., Akhter, J.N., Mazid, M.A., Halder, C.G., 1993. First record of induced breeding of Thai Panyas, *Pangasius sutchi* (Fowler) in Bangladesh. J. Inland Fish. Soc. India 25 (2), 26–30.

Ruzzante, D.E., Hansen, M.M., Meldrup, D., Ebert, K.M., 2004. Stocking impact and migration pattern in an anadromous brown trout (*Salmotrutta*) complex: where have all the stocked spawning sea trout gone? Mol. Ecol. 13, 1433–1445.

Ryman, N., Laikre, L., 1991. Effects of supportive breeding on the genetically effective population size. Conserv. Biol. 5, 325–329.

Saidin, T., Othman, A.F., 1986. Induced spawning of Pangasius sutchi (Fowler) using an analog of luteinising releasing hormone and homoplastic pituitary extract. In: Procedings of the First Asian Fisheries Forum, pp. 687–688.

Sun, X., Liang, L., 2004. A genetic linkage map of Common carp, Cyprinus carpio L. and mapping of a locusassociatedwith cold tolerence. Aquaculture 238, 165–172.

Talwar, P.K., Jhingran, A.G., 1991. *Inland Fishes of* India *and Adjacent Countries*, Vol. I. Oxford & IBH Publishing Co. Pvt. Ltd, New Delhi/Calcutta, pp. 173–174.

Thomas et al., 2003. Fluorescence In Situ Hybridization (FISH)—Application Guide.

Vandeputte, M., 2003. Selective breeding of quantitative traits in the common carp (*Cyprinus carpio* L.): bases, results and prospects. Aquat. Liv. Res. 16, 399–407.

Vandeputte, M., Kocour, M., Mauger, S., Dupont-Nivet, M., De Guerry, D., Rodina, M., Gela, D., Vallod, D., Chevassus, B., Linhart, O., 2004. Heritability estimates for growth-related traits using microsatellite parentage assignment in juvenile common carp (*Cyprinus carpio* L.). Aquaculture 235, 223–236.

Wu, T., Yang, H., Dong, Z., Xia, D., Shi, Y., Ji, X., et al., 1994. The integration and expression of human growth gene in blunt snout bream and common carp. J. Fish. China Shuichan Xuebao 18, 284–289.

8

SEED RESOURCES AND SUPPLY CHAIN IN SOME ASIAN COUNTRIES

More than 75 species in total are in use both for breeding and culture practices in Asian countries (FAO, 2005b) of which China alone has more than 50 species in the freshwater sector. Of these, 30 freshwater fish and prawn species are native to the Mekong Basin, with a diverse collection of both exotic and indigenous species

Induced Fish Breeding. DOI: http://dx.doi.org/10.1016/B978-0-12-801774-6.00008-0

(Phillips, 2002). Fish seed for stocking are either collected from the wild, hatcheries, or are imported from other countries, while an extra source is bundhs. Bundhs are specialized ponds where seeds are produced through sympathetic breeding and the quality is like that of native fish. The species mainly considered for seed production (Table 8.1) include silver carp, bighead carp, grass carp, silver barb, common/mirror carp, Indian major carp, Thai pangus, Thai koi, Nile tilapia, pacu, Thai magur, African magur, and some important minor carp along with genetically improved farmed tilapia (GIFT). Some are bred for conservation purposes, e.g., Mahseer and black carp in Bangladesh, and others for culture. As composite farming is the choice of farmers so there is a demand for a mixture of recommended species, and this may be the cause of fish breeders adopting mixed spawning. Until the early to mid-1970s, riverine collection was the one source except bundh breeding in India and accounted for 91.67% and 97.5% of the total fish seed production during the mid-1960s in India and the 1980s in Bangladesh, respectively. The vast stretches of low-lying lands bounded by embankments that are filled with run-off from extensive catchment areas during the monsoon have been traditionally used in India as "natural hatcheries," called bundhs, to produce wild seed of carp. The sudden influx of rainwater into these systems provides a stimulus for the fish to spawn (Mondal et al., 2005). This natural way of induction to spawn accounted for a major portion of fish seed during the 1960–1980s in India. In Bangladesh, carp spawn and fry were collected during the monsoon season from different river systems, while collected fertilized eggs were incubated in earthen pits in the riverbanks for hatching. This method yielded low hatching and higher mortality rates.

Wild seed registered a declining trend due to pollution, habitat loss, and a lot of anthropogenic activities in most Asian countries. The development of captive breeding through the on-field standardization of induced breeding technology not only created a revolution in the seed production sector but shouldered the responsibility to supply 70–80% in most Asian countries. This success led to the development of mass room hatcheries in all the countries, mostly in the private sector. With the progress of time, dependency on wild seed reduced to 5% in India and 0.45% in Bangladesh, though still the wild habitat acts as a source of some species.

In the changed senario wild resource still acts as a significant supply chain in Cambodia (26%) and China (20%) elucidating a lower pollution load and other anomalies than encountered by the other Asian countries. This includes fingerlings of snakehead (*Channa micropeltes* and *Channa striata*), pangasiid catfish (*Pangasianodon hypophthalmus*, *Pangasius conchophilus*, *Pangasius larnaudiei*, and *Pangasius bocourti*), catfish (*Hemibagrus wyckioides*), and barbs (*Barbonymus gonionotus* and *Barbonymus altus*) (Table 8.2). Among

Table 8.1 Fish Species Used in Seed Production for Freshwater Aquaculture in 10 Countries in Asia (FAO Report , 2005a)

Country	Aquaculture Species
Bangladesh	*Catla catla, Labeo rohita, Cirrhinus cirrhosus, Cirrhinus ariza, Labeo calbasu, Labeo bata, Labeo gonious, Puntius sarana, Hypophthalmicthys molitrix, Cyprinus carpio, Ctenopharyngodon idella, Aristichthys nobilis, Barbonymus gonionotus, Pangasius sutchi, Oreochromis niloticus, Anabas testudineus, Labeo bata, Labeo gonia, Clarias batrachus, Heteropneustes fossilis, Ompok pabda, Mystus cavasius, Mylopharyngodon piceus, Tor putitora, O. mossambicus × O. niloticus*
Cambodia	*Barbonymus gonionotus, Hypophthalmichthys molitrix, Cyprinus carpio, Oreochromis niloticus, Cirrhinus cirrhosus, Pangasianodon hypophthalmus, Leptobarbus hoevenii, Barbonymus altus, Clarias macrocephalus, Trichogaster pectoralis, Aristichthys nobilis, Labeo rohita, Channa micropeltes* and *Channa striata, Pangasius conchophilus, P. larnaudiei, P. bocourti, Hemibagrus wyckioides*
China	*Ctenopharyngodon idellus, Hypophthalmichthys molitrix, Aristychthys nobilis, Cyprinus carpio, Oreochromis niloticus, Mylopharyngodon piceus, Barbodes goniotus, Cirrhinus molitorella, Labeo rohita, Catla catla, Cirrhinus mrigala, Clarias* sp., *Channa* sp.
India	*Cyprinus carpio, Ctenopharyngodon idellus, Hypophthalmichthys molitrix, Aristychthys nobilis, Labeo rohita, Catla catla, Cirrhinus mrigala*
Indonesia	*Cyprinus carpio, Oreochromis niloticus, Osphronemus gouramy, Clarias batrachus, Clarias garipienus, Pangasius suchi, Pangasius jambal, Puntius gonionotus, Osteochillus hasselti, Collosoma* sp., *Macrones* sp., *Leptobarbus hoeveni, Rana catesbiana, Cherax* sp., *Oxyeleotris marmoratus, Notopterus chitala, Channa* sp., *Anabas testudeneus, Trionyx* sp.
Pakistan	*Salmo gairdneri, Salmo gairneri, Cirrhinus mrigala, Labeo rohita, Catla catla, Tor putitora, Hypophthalmichthys molitrix, Ctenopharyngodon idella, Cyprinus carpio, Aristichthys nobilis*
Philippines	*Oreochromis niloticus*
Sri Lanka	*Ctenopharyngodon idellus, Hypophthalmichthys molitrix, Aristychthys nobilis, Labeo rohita, Catla catla, Cirrhinus mirigala, Chanos chanos*
Thiland	*Clarias macrocephalus* x *C. gariepinus, Oreochromis niloticus, Barbodes gonionotus, Trichogaster pectoralis, Macrobrachium rosenbergii, Pangasianodon hypophthalmus, Channa striata, Cyprinus carpio, Trionyx sinensis*
Vietnam	*Cyprinus carpio, Ctenopharyngodon idellus, Hypophthalmichthys molitrix, Hypophthalmichthys harmandi, Aristychthys nobilis, Labeo rohita, Catla catla; Cirrhinus cirrhosus, Cirrhinus molitorella, Mylopharyngodon piceus, Spinibarbus denticulatus, Carassius auratus, Barbodes goniotus, Oreochromis niloticus, Pangasius hypopthalmus, Clarias macrocephalus × C. gariepinus, Notopterus notopterus, Anabas testudineus, Oxyeleotris marmoratus, Ophiocephalus micropeltes*

the seeds collected nowadays, wild carnivorous species account for 95%, with an annual catch amounting to 20 million fingerlings in Cambodia. In India, for last few years, there has been a craze for hatchery owners to import catfish from various Asian, European,

Table 8.2 Freshwater Fish Species Used in Hatchery Production of Spawn or Hatchlings in Some South Asian Countries (FAO Report, 2005b)

Species	Hatchery-Induced Breeding	Hatchery Natural Breeding	Gene Banking
Anabas testudineus	+		
Aristichthys nobilis	+	+	L
Barbonymus gonionotus	+	+	
Carassius carassius	+		
Catla catla	+		
Cirrhinus ariza	+		
Cirrhinus cirrhosus/mrigala	+		
Clarias batrachus	+		
Clarias macrocephalus × *C. gariepinus*	+		
Ctenopharyngodon idella	+	+	L,C
Cyprinus carpio	+	+	C
Heteropnuestes fossilis	+		
Mystus cavacius	+		
Ompok pabda	+		
Anabas testudineus	+		
Labeo rohita	+		
Ompok pabda	+		
Oreochromis niloticus	+	+	
Hypophthalmichthys molitrix	+	+	C, L
Ictalurus punctatus	+		
Macrobrachium rosenbergii	+	+	
Mylopharyngodon piceus	+		
Mystus cavacius	+		
Pangasius hypopthalmus	+		C
Pangasius sutchi	+		
Puntius sarana	+		

C, cryopreservation; L, live gene bank.

and American countries and sell their seed at high price after standardization of breeding in captivity. In this endeavor, leading major carp hatcheries have been converted into exotic cat fish hatcheries—a profit-making approach thought to create further negative consequences on the major carp hatchery sector in the absence of any control and follow-up action. Wild fry collections of the euryhaline species *Chanos chanos* were practiced in Sri Lanka until the late 1980s for both freshwater and brackishwater aquaculture. Despite the estimated collection potential of 600 million fry/annum (Ramanathan, 1969), only 1.3 million were collected between 1979 and 1983 (unpublished data; Thyaparan and Chakrabarty, 1984). The milkfish fry collection program was abandoned in 1989 as a result of the withdrawal of state patronage to inland fisheries in 1990. Therefore, no estimation could be made on the current status of this wild seed resource.

Fish seed are also imported by different countries in the face of: (1) inconsistent supply from local hatcheries of certain species in demand, (2) lower price, and (3) rapidly depleting wild seed resources. During the 1980s and 1990s, Cambodia used to export billions of *P. hypophthalmus* fry and millions of *P. bocourti* fingerlings collected from the Mekong Basin to neighboring countries. Now Cambodia imports fish seed of six exotic species and three indigenous species in the range of 60 million fingerling per annum, representing 56% of its fingerling requirements, from Vietnam and Thailand. This indicates the rate at which natural populations of these species have declined. It is believed that the imports of hybrid catfish will increase as a result of the recent ban on giant snakehead cage culture operation due to its dependence on small wild fish for dietary nutrient inputs (So et al., 2005). As a consequence, escapes of this hybrid catfish may bring negative impacts on the aquatic environment. Ing Kimleang (2004) reported that many unlicensed traders imported an estimate of about 200 million fingerlings from Vietnam (mostly) and Thailand to Cambodia for diverse culture practice, indicating the transboundary movement of aquatic animals is largely unregulated. There are implications for disease transmission and genetic pollution. The possible reason for importing fish seed from neighbouring countries for sustenance of aquaculture in Cambodia are:

- Local hatcheries are unable to meet the demand of fish seed for aquaculture;
- Inconsistent supply;
- Cheaper prices of imported fish seed than local hatcheries;
- Market potential of fish seed in Cambodia;
- Depleted wild fishery resources.

With the advent of induced breeding technology of Indian and Chinese major carp, it became possible to obtain quality seed of carp for aquaculture. This resulted in an increased reliance on induced breeding for obtaining quality fish seed. At present, induced breeding accounts for most of the seed produced by many species throughout Asia. Although artificial propagation is the main means of seed production, seed collected from the wild is mainly used for raising quality brood stocks. Brood stocks, used for artificial propagation, are usually raised in captivity by collecting seeds from the wild or from breeding centers where good natural stocks are maintained. Private-sector hatcheries and nurseries, particularly hatcheries and fry nurseries composed of wild collections, play a significant source of fish seed and far exceed the public sector contribution.

8.1 Seed Management

8.1.1 Hatchery Management

Hatchery management mainly consider multispecies activities when different species (both indigenous and exotic) are maintained in the same pond, except by large-hatchery owners. As we know brood stock is the primary prerequisite to satiate the principle of induced breeding, i.e., production of quality seed but mixed culture may not satisfy the species-specific management and the result is incomplete maturity and fecundity, failure of reproduction, along with reduced fertilization and hatching. The fish breeders and hatchery owners are not at all conversant with these aspects of management practices which play a crucial role in quality seed production, on the contrary they adopt a technique which is convenient, less hazardous, and economic. The scientific management practices are generally categorized into a two-phase program:

1. Prespawning and
2. Postspawning management.

Prespawning management includes brood stock selection, procurement, stocking, fertilization, feeding, routine check-up in relation to maturity and health, acclimatization, spawning, and hatching; while postspawning includes seed maintenance, water quality management, brood restocking, selection and transfer/transport of spawn, rearing of spawn, health management, assessment of condition, selection and risk assessment, and documentation and record-keeping. Since selective breeding and hybridization programs of pedigree fish is seldom practiced in most hatcheries so procurement of brood stock is undertaken either by collecting individuals from the wild or from neighboring farms or hatcheries, or by developing new brood stocks from the early stage. The negative aspect of selecting fish from hatchery-raised fish is that only undersized and residual fingerlings are considered for raising brood stock since larger and fast-growing fingerlings are sold first. Therefore, scientific mass selection procedure, as advocated, which consists of rearing fast-growing better-quality fry and fingerlings with correct body shape, color, and free from external deformations is not practiced. Again, selection of females and males from two different family lines, to reduce the chances of inbreeding, is not practiced at all and even the fish breeders are not appraised of its advantages and disadvantages. The good practice of transferring spawners after spawning to a resting pond without mixing with ripe males and females is ignored not only in small and marginal-scale hatcheries but also in large hatcheries. Efforts are initiated to improvise brood stock through specific genetic improvement programs (GIFT, GET EXCEL, GST) to select for desirable traits or simply selected from stocks that show better traits (color, growth) or are free from, or suspected to be resistant or tolerant to, specific conditions or pathogens.

The development of GIFT tilapia has clearly demonstrated that rapid genetic improvement of farmed fish through selective breeding is possible. In Bangladesh, the Philippines, Thailand, and Vietnam, national breeding programs and related tilapia genetic research are based mainly or exclusively on GIFT or GIFT-derived strains through selective breeding. GIFT tilapia is now extensively used as a reference point for research on the development of tilapia farming in many Asian countries including China and Indonesia. Methods used for the development of GIFT tilapia are being used for genetic improvement of other species such as silver barb in Bangladesh and Vietnam, rohu in Bangladesh and India, mrigal in Vietnam, and blunt snout bream (*Megalobrama amblycephala*) in China. The issues related to brood stock management in captivity experienced by the hatchery sector include inbreeding, genetic drift, introgressive hybridization, and unconscious selection. Ingthamjitr (1997) reported that hybrid *Clarias* catfish farmers in central Thailand, instead of using fairly standard management practices, had suffered a decline in production of good-quality seed which was subsequently found to be related to the quality and genetic management of the brood stock.

Brood fish of fecund species for which relatively few fish are used to spawn may be particularly prone to mismanagement (Little et al., 2002). Lack of regular introduction of fresh fish germplasm, with timed periodicities from natural sources or from distant hatcheries, was believed to be the main reason for inbreeding depression and reported to have occurred in some carp hatcheries in India (Eknath and Doyle, 1985). Over short timescales, it is unlikely that inbreeding is a major contributor (Mair, 2002). The common and wide belief of inbreeding as the cause of poor-quality seed often overlooked the husbandry management and environmental factors that might be responsible for poor-quality seed. Moreover, the current schemes to upgrade stocks with wild or improved fish (e.g., GIFT) may be unsustainable unless problems are better understood and improved management strategies developed (Little, 1998). To avoid potential problems related to genetics, poor growth, and survival due to inbreeding pressure, details of the different families/lines or origin of the domestic stocks (foreign or native) must be taken into account along with the record of performance and development data for the families or lines under a range of identical environmental conditions. The selection protocol used is also important, i.e., whether the stocks were selected from ponds with better performance or survivors following a disease outbreak and the exact timing of the selection procedures. Large hatcheries as well as government hatcheries may have a system of recordkeeping.

Most of the hatcheries, however, do not maintain and are not appraised of maintaining any records related to pedigrees or sources. The fish breeders and hatchery operators should be made aware of the importance of recordkeeping as well as the criteria for selection and purchase of broods from other sources. It is realized now that some leading and large resourceful hatchery owners should be entrusted with selling broods with proper certification. Mouth-brooding tilapia females undergo a nonfeeding stage during buccal incubation which lasts for 10–13 days. Female tilapia need to be fed actively with quality feeds after this period to regain body condition lost during incubation and to obtain energy to support further reproductive activity. Most small-scale hatcheries may not meet this feed requirement due to the high cost. Therefore, alternative feed sources should be made available for small-scale farmers or an appropriate formulation for farmer-made aqua-feeds. One striking area that is lacking in brood stock management is the procedures for brood stock quarantine. Quarantine facilities are essentially a closed holding area where brood fish are kept in individual tanks until the results of screening for known diseases or disorders are known. Such facilities can be afforded by most large-scale and government hatcheries but not by small-scale or farmer-operated hatcheries. Quarantine holding facilities should ideally be kept some distance from the hatchery. In cases where this is not possible, measures should be taken to ensure that there will be no contamination from the holding facility to the other production areas. The success of the fish seed nursing is critical as the fish seed production of a country is measured by the quality output from fry nurseries. Nursery operation can be carried out in two ways: (1) single-stage operations where hatchlings are raised until fingerling stage and (2) two-stage operations such as: (i) raising hatchlings to fry and (ii) raising fry to fingerlings. In a two-stage operation hatchlings are stocked at higher densities (e.g., 6 million carp hatchlings/ha in Bangladesh) than that in a single-stage operation (1–2 million hatchlings/ha) where fry are thinned out and stocked at lower stocking densities (0.2–0.3 million/ha) for rearing up to fingerling size. Sometimes further rearing stages are included by splitting fry rearing into early and late fry rearing. Nursing operations tend to concentrate near centers of trading. Most nursery operators learned the nursing techniques from their neighbors. The types of species used vary within the country depending on the region. The main species used in nursery operations in the southeast region of Vietnam are common carp and grass carp, while in the southwest region no single species dominates, although silver barb and giant gourami are more common (AIT, 2000). Nursery operators commonly purchase hatchlings from private hatcheries or use seed produced from their own hatcheries. Experienced nursery operators purchase hatchlings from government hatcheries since they believe that hatchling quality is higher in government hatcheries (AIT, 2000). Small-scale hatcheries and nurseries are usually located near the operators' homestead, enabling close monitoring and employing family labor. This proximity to home allows greater participation of women in hatchery activities. Nursery operations are based on stocking of a single species of hatchlings in fertilized earthen ponds at high stocking densities. Fish are raised as a batch for a few weeks or months before harvesting. Government hatcheries which operate nurseries are expected to remain an important source of seed, alongside the growing number of private and corporate nursery operators. Government hatcheries will mostly remain as adjunct suppliers and, to a certain extent, a competitor to private seed producers. The continued presence of the public sector in the seed market will benefit small farmers through the sector's promotion of hatchery and nursery management and other developments to various parts of the country. Although seed supply in most remote areas is still problematic, efforts by the public sector contributed to increased access and wider choice of seed resources for the benefit of small-scale farmers.

8.2 Seed Production Facilities and Seed Technology

During transformation from wild collection to hatchery production, hatcheries were used as a facility for hatching of fish eggs collected from the wild. Over the years, the development and refinement of the technique of induced breeding of carp has been improved. Hence, more emphasis was given to the proper use of hatcheries for large-scale production of fry. Artificial fish breeding techniques and low-cost hatchery designs have been successfully adopted during the past three decades. Breeding techniques for Indian major carp (catla, rohu, and mrigal) and Chinese carp were first developed mainly by public-sector hatcheries from donor-assisted projects. These public-sector hatcheries served as demonstration units. The technological progress in induced breeding of Indian and Chinese major carp was extended to a number of species such as silver carp, bighead carp, grass carp, silver barb, Indian major carp, common carp, Thai pangus, and GIFT tilapia (Table 8.3). Other species such as mahseer are being bred for conservation purposes in Bangladesh and India.

The evolution of carp hatchery systems in India was reviewed by Dwivedi and Zaidi (1983). These hatchery facilities range from simple earthen pits to pots, cement tanks to hapas, transparent glass and polyethylene jars to galvanized jars, tanks, and Chinese-style circular tanks. Chinese-style circular spawning and incubation systems introduced from China or its modifications dominate in the region (Vietnam, Sri Lanka, Bangladesh, Pakistan). The low-water head and large flow rates required by these systems are more suitable for

Table 8.3 Commonly Used Hormones in Induced Breeding

Species	Hormone Extracts	Synthetic Agents	Country
Ctenopharyngodon idellus, Hypophthalmichthys molitrix, Aristychthys nobilis, Labeo rohita, Catla catla, Cirrhinus mrigala	Sufrefact, human chorionic gonadotropin, pituitary gland	Ovaprim, LHRH analogue, Surfrefact	Sri Lanka
Ctenopharyngodon idellus, Hypophthalmichthys molitrix, Aristychthys nobilis, Labeo rohita, Catla catla, Cirrhinus cirrhosus, Labeo calbasu, Barbonymus gonionotus, Puntius sarana, Pangasius sutchi, Clarias batrachus, Heteropnuestes fossilis, Ompok pabda, Anabas testudineus	Pituitary gland, human chorionic gonadotropic hormone	Ovaprim, Profasi, Pregnyl, LRHA	Bangladesh
Hypophthalmichthys molitrix, Cyprinus carpio, Barbonymus gonionotus, Clarias macrocephalus × *C. gariepinus, Labeo rohita, Catla catla*	Pituitary gland, luteinizing hormone-releasing hormone, human chorionic gonadotropic		Vietnam
Major carp	Pituitary gland, human chorionic gonadotropic	Ovaprim, Ovatide	India
Ctenopharyngodon idellus, Hypophthalmichthys molitrix, Aristychthys nobilis, Labeo rohita, Catla catla		Ovaparim, LHRH analogue	Pakistan

areas not experiencing water limitations. However, these hatcheries work well for breeding of Chinese and Indian carp, *Spinibarbus*, and silver barb. Hapas are mainly used for spawning of common carp and sometimes tilapia. Induced spawning and fertilization of stinging catfish (*Heteropneustes fossilis*) is being conducted in hapas. Several hormone extracts as well as some synthetics are used in induced breeding technology (Table 8.4). The most commonly used hormones are pituitary gland extract (PG) and luteinizing hormone-releasing hormone (LHRH). Due to dire shortage of brood fish, multiple breeding is adopted in most of the hatcheries in Asian countries, including India. The same carp are used as broods 2–3 times in a season at a gap of one to one and a half months. Inappropriate use of hormones has been implicated to result in poor-quality seed early in the season. As reported by the fish breeders, the price of fish seed produced using only LH-RH is lower than that using PG, reflecting the view of farmers who claim that LH-RH-produced seed are of poor quality (AIT, 2000). Fertilization of eggs is carried in three ways: (1) natural fertilization, (2) artificial fertilization by stripping and "dry" fertilization, or (3) by both means. In the case of natural fertilization, induced males release sperm at the same time that females release the eggs in the same tank for fertilization to take place. Fertilized eggs are collected and placed in the incubator for hatching. In the case of common carp or mirror carp breeding, a sufficient quantity of aquatic grasses is placed in the breeding tanks prior to spawning for use in the collection of the sticky fertilized eggs.

Artificial fertilization in fish is achieved by means of the following: (1) stripping sperm from the induced male and manually fertilizing eggs released by induced female (carps, pangus fish), (2) stripping both eggs and milt without using inducing agents (*Oreochromis niloticus*), and (3) fertilizing manually, or by stripping eggs from the induced female and milt from macerated testes of sacrificed males (catfish). In the case of the mouth-brooding tilapia (*Oreochromis* spp.), sometimes the fertilized eggs are taken from the mouth of the mother for incubation. To enhance hatching success, the sticky eggs of certain species (e.g., common carp, mirror carp, catfish) are washed to remove stickiness or spread evenly in the incubation tray. Fresh milk or clay-dissolved water are used to remove the stickiness. Small-scale hatchery operators prefer to breed fast-growing fish with low-cost diets (e.g., common carp, silver carp, mrigal, and tilapia). On-farm research on breeding of fish attaining early maturity, with small brood stock such as Java barb (*B. gonionotus*), mad barb (*Leptobarbus hoevenii*), snakeskin gourami (*Trichogaster pectoralis*) and *Osteochilus melanopleura* are carried out in Cambodia. High fecundity and high fertilization, hatching, and survival rates have been achieved for Java barb.

Table 8.4 Pituitary Extract and Synthetic Hormone Used in Fish Breeding

Species	Hormone Extracts	Synthetic Agents	Country
Cat fish: *Barbonymus gonionotus, Puntius sarana, Pangasius sutchi,* olive barb; *Clarias batrachus; Clarias gariepinus,* African catfish, *Heteropnuestes fossilis,* Asian catfish, *Ompok pabda, Anabas testudineus,* Butterfly catfish; *Mystus cavasius,* Gangetic mystus; *Anabas testudineus,* Climbing perch	Pituitary gland, human chorionic gonadotropic hormone, Steroid hormones Deoxycorticosterone acetate (DOCA) and cortisone effectively stimulate (dosage 50 mg/kg of fish) ovulation in *Heteropneustes fossilis* (Goswamy and Sunderraj, 1971)	Ovaprim, Profasi, Pregnyl, LRHA	Sri Lanka
Aquarium fish Cold fish, trout, and pike *Hypophthalmichthys molitrix, Cyprinus carpio, Barbonymus gonionotus, Clarias macrocephalus × C. gariepinus, Labeo rohita, Catla catla*	17 α -hydroxy-20B dihydroprogesterone (17 α -20BDP) is useful to induce (Jalabert, 1973) Pituitary gland, luteinizing hormone-releasing hormone, human chorionic gonadotropic		Bangladesh Vietnam
Major carp *Labeo rohita,* rohu, *Catla,* catla: *Cirrhinus cirrhosus,* mrigal: *Labeo calbasu,* calbasu: *Cirrhinus ariza,* Reba carp; *Labeo bata,* Bata; *Labeo gonious, Kurio labeo*	Pituitary gland, human chorionic gonadotropic Clomiphene, an analogue of the synthetic nonsteroidal estrogen chlorotrianisene possess antiestrogenic effects in teleosts	Ovaprim, Ovatide, LHRH analogue, Surrefact	India, Bangladesh, Pakistan, Vietnam
Exotic Carp *Ctenopharyngodon idella,* grass carp; *Hypophthalmichthys molitrix,* silver carp *Aristychthys nobilis,* bighead carp; *Mylopharyngodon pices* black carp Nile tilapia, *O. niloticus,* GIFT tilapia, *O. mossambicus × O. niloticus;* red tilapia, *Anabas testudineus;* climbing perch (Thai koi)	Surrefact, human chorionic gonadotropin, pituitary gland	Ovaprim, LRHH Analogue, Surrefact	India, Bangladesh, Pakistan, Vietnam

8.2.1 Gene Banking

Interest in the development of gene banks to improve the genetic quality of brood fish and thus the seed quality emerged as an evolutionary technology which opened a new window for accelerated development in the aquaculture sector. Large collection of crop varieties and carefully maintained livestock breeding nuclei and cryopreserved sperm and embryos, commonly called gene banks, are the basis of most of the world's plant and livestock breeding programs and related research. By comparison, fish gene banks are rare and supported inadequately, especially in tropical developing countries (Brian et al., 1998). One of the most valuable fish gene banks in Asia is the Nile tilapia brood stock assembled for the development of GIFT, together with the GIFT synthetic base population and subsequent generations of selectively bred GIFT in the Philippines at the National Freshwater Fisheries Training Centre (NFFTC). The descendants of these fish remain available from this gene bank for national, regional, and international breeding purposes.

Considerable progress has been observed in Bangladesh with respect to gene banking. From 2002 to 2003, the Department of Fisheries of Bangladesh initiated the establishment of 12 fish brood stock banks in the Government Fish Seed Multiplication Farms with target production of 110 tonnes of genetically improved brood of Chinese carp, 1800 kg spawn, and 0.5 million fingerlings. The fry for the brood stock banks were collected from different rivers. To date, 85 tonnes of brood have been produced and distributed to public and private hatcheries under the newly formulated fisheries policy. Another 20 brood banks have been established in 20 Fish Seed Multiplication Farms and one Fish Breeding and Training Centre under the Fourth Fisheries Project. Moreover, an NGO (BRAC) has established one carp brood bank in the hatchery and is planning to set up another. These brood banks provide the necessary training for other government and private hatchery operators on brood stock management. Although a cryogenic gene bank has not been established as yet, research is in progress in the cryopreservation of sperm of Indian major carp (*Catla catla, Labeo rohita, Cirrhinus cirrhosus*) and Chinese carp (*Cyprinus carpio, Ctenopharyngodon idella, Hypophthalmichthys molitrix, Aristichthys nobilis, B. gonionotus*).

8.2.2 Fish Seed Nursing

Fish seed nursing is either carried out in earthen shallow ponds (for carp), cemented tanks (for catfish), or hapas and net cages (for tilapia). Development of the hapa fry nursing has been recognized as a means of supporting the fingerling requirements in remote areas lacking good infrastructure. Both seasonal and perennial ponds are used as nurseries in Bangladesh to cater to the fingerling requirements in remote areas. Adherence to standard pond preparation procedures, similar to brood stock pond preparation, before stocking the ponds with postlarvae or fry is a common practice. Pond preparation is crucial for getting algae, which is a prerequisite for stocking of carp spawn and fry. Prior to stocking of fingerlings, ponds are prepared by dewatering or by using toxins to remove unwanted fish. Aquatic weeds are manually removed. Liming and fertilization procedures are more or less the same as those of brood stock pond preparation. Along with some supplemental feeding, farmers are mostly dependent on natural food produced in the ponds for fingerling rearing. To grow a sufficient amount of planktonic food, urea, TSP, and cow dung are applied in high doses. Preparation of compost at the corners of the ponds is common. As supplementary feed, mustard oil cake, rice bran, wheat bran, and sometimes fish meal are used.

8.3 Fish Seed Quality

8.3.1 Fish Seed Quality Assurance

Poor-quality seed, perceived as a major constraint to the expansion of fish culture, can have a deleterious effect on fish production and brood stock development. Little, Satapornvanit and Edwards (2002) have emphasized the importance of freshwater fish seed quality in Asia and suggested criteria for selecting good-quality seed for aquaculture. Some countries have adopted several approaches to ensure fish seed quality. China has established an institutional approach to ensure fish seed quality. As per the "Aquatic Seed Management" definition of the Ministry of Agriculture, fish breeders should obtain the brood stock from one of the centers established under the National Aquatic Bred and Wild Seed System (NABWSS), which includes genetic breeding center (GBC), wild variety collection center (WVCC), wild/bred variety amplifier (WBVA), exotic species centers (ESC), and seed quality inspection centers (SQIC). A certification process has also been adopted but limited to the authorization of the release of genetically improved varieties. Under this certification process the National Certification Committee of Aquatic Wild and Bred Varieties (NCCA-WBV) has authorized the release of 32 strains, including 16 selective and 16 crossbreeding strains. The success of such a public institutional approach suffers from inability to meet the demand for brood stock and more emphasis on licensing than quality assessment. Criteria and standards are available to assess seed and brood stock quality. Quality criteria are largely based on age and uniformity of size in terms of weight and length (Indonesia, Thailand), growth performance, survival and percentage of deformities (Vietnam), and body shape and behavior (Sri Lanka). Some farmers have experienced sudden spawn deaths and the occurrence of deformed larvae/fry, more commonly for spawn produced during the late breeding season, thus farmers tend to avoid breeding during the late spawning season. Further investigations are required to select and establish better criteria for quality assurance.

In Bangladesh and Indonesia, fish seed quality criteria have been adopted as national standards for eight species (common carp, Nile tilapia, Siamese fighting fish, Catfish, walking catfish, gouramy, giant freshwater prawn, and bullfrog). The quality of fish seed for sex-reversed tilapia (SRT) and genetically male tilapia (GMT) are often assessed by the occurrence of any breeding in the production stock from unwanted females. Fish seed quality assurance through health management is widely used as disease is considered as one of the most important problematic factors in the seed industry. Parasitic diseases in nurseries are one of the most important limiting factors for growth and survival of fry and fingerlings. In many Asian countries, severe mortalities among carp fry have been reported, caused by different parasites such as *Ichthyopthirius* sp. (ich disease), *Trichodina* spp., *Ichthyobodo* spp., *Lernaea* spp., *Myxobolus* spp., and *Dactylogyrus* spp., *Myxobolus* and *Henneguya* affect the gills and have caused heavy mortalities in *C. catla*. The most commonly adopted approach is periodic checks on the health status of fish seed and carrying out recommended treatments (Sri Lanka, Thailand, Bangladesh, India). During an outbreak of a disease in hatcheries and nurseries, farmers used different treatments, such as chemicals and antibiotics, water exchange, and manipulation of feeding and fertilization. Prophylactic treatment is a common practice. Health management practices are often considered as part of good husbandry (Sri Lanka, Pakistan).

There are no specific/unique procedures or standards developed for maintaining hygienic conditions in fish hatcheries, but most hatcheries take precautionary measures before and during fry production. As a part of these precautions, disinfection of facilities and materials in use is commonly practiced. The FAO Code of Conduct for Responsible Fisheries (CCRF):
- Avoid inbreeding;
- Maintain stock integrity by not hybridizing different stocks, strains, or species;
- Minimize transfer of genetically different stocks;
- Periodically assess their genetic diversity (i.e., by laboratory genetic analysis).

Inbreeding, interspecific hybridization, negative selection of broods, and improper brood stock management are reported as common phenomena in hatcheries, particularly in small-scale hatcheries.

Hussain and Mazid (1997) reported reduced growth, physical deformities, diseases, and high mortality in hatchery-produced carp seed and they have identified improper management of brood stock, unconscious negative selection of broods, unplanned hybridization, and inbreeding as probable reasons behind these reduced performances. Recent studies revealed high rates of inbreeding and interspecific hybridization in both endemic and exotic carp (Simonsen et al., 2005; Simonsen et al., 2004; Alam et al., 2002). These factors result in low growth rate, high mortality, deformities of fish seed, and less fecundity. The measures taken to ensure seed quality are selective breeding, ease of inbreeding pressure, genetic control and manipulation, and gene banking. The deterioration of genetic quality in several cultured fish species in Vietnam has been regarded as very striking since the late 1970s. Selective breeding programs are available for freshwater fish such as common carp (Tien, 1993), tilapia (Dan et al., 2001), Mekong striped catfish (Hao et al., 2004), grass carp and mrigal (Tuan et al., 2005), and tilapia. Inbreeding pressure is high in carp; hatchery operators do not (1) maintain an effective population size and (2) exchange brood fish between hatcheries. Poor performance of the resultant seed has been linked to inbreeding of carp in India (Eknath and Doyle, 1985). A communal or mixed spawning system for major carp in West Bengal is being practiced and is known to produce approximately 10% hybrids. This technique may lead to loss of genetic purity of important major carp.

Fish farmers often complain about poor growth of fish procured from certain hatcheries, particularly small-scale or farmer-operated hatcheries. They feel that such fish do not reach marketable size within the stipulated period. This is also attributed to inbreeding. It is believed that small-scale hatchery operations, particularly farmer-operated hatcheries, can rapidly give rise to deterioration of brood stock quality due to their limited capacity to maintain minimum effective population size of brood stock. Small-scale farmer hatcheries usually maintain multispecies brood stock in one or two ponds with excessive stocking densities. The brood stock ponds were usually either underfed or fed with low-quality feed. Competition between fish species may also limit the potential of each stocked fish species in terms of maturation, fecundity, fertilization, hatching success, and survival rates. It is reported that these hatcheries rarely recruit new broods from outside. If they do, the recruitment is from the subsequent generations of the same parent stock without any inflowing of new genetic material. In many hatcheries, the common practice of using the same brooders more than once in a breeding season causes deterioration of larval quality, mortality, and larval deformity. Hatcheries are more concerned with quantity rather than the quality of fish seed; production does not follow any selection procedure, thus, high mortality rates, poor growth, and high susceptibility to disease and other parasitic infections are common occurrences. Several initiatives have been taken to preserve the genetic quality of brood stock to assure high-quality seed. Mass selection is the most ancient and simplest method to improve the seed of species used in fish culture. Mass selection of common carp has been carried out in Vietnam where a 33% growth increase was achieved at the fifth generation (Tien et al., 2001). This slow progress is due to a low degree of inheritance in quantitative traits. To enhance the quality, mass selection was replaced by family selection. To improve seed quality, measures have been taken such as establishment of brood banks (Bangladesh) or gene banks (Philippines). Recently, emphasis has been placed on increasing awareness among farmers and hatchery operators concerning genetic issues in fish breeding to improve brood stock management practices. Lack of records on breeding individuals, mixed spawning of fish species, and lack of knowledge or awareness on minimum effective population size have been identified as bottlenecks to improve genetic management in small-scale or farmer-operate hatcheries.

8.4 Institutional Support and Seed Certification

8.4.1 Fish Seed Certification

A proper fish seed certification process does not exist in many countries in the region. Table 8.5 shows the kind of certification existing in six countries. In Bangladesh, the Ministry of Fisheries and Livestock has drafted the Law for Fish and Shrimp Hatchery 2005 known as "Matsha and Chingri Hatchery Ayen, 2005," which includes requirements for registration of hatcheries and the rules for fish and shrimp hatchery operation with elaboration on aspects of hatchery operation such as physical infrastructure/facilities of hatchery, ponds, selection of brood fish for breeding, source of selected brood fish, environment, etc. Similar certification processes on hatchery management and operation exist in Indonesia. These initiatives will help improved the quality of fish seed produced.

8.4.2 Policy and Legislation

Fisheries and aquaculture are governed by fisheries policy. In the recent past, recognizing aquaculture finds its place in government policies. For example, in Bangladesh, the Fisheries Policy adopted in 1998 included policy for freshwater aquaculture (Section 6.0 of the

Table 8.5 Mode of Fish Seed Certification

Country	Authority	Purpose
Bangladesh	National Committee	Preparing a policy for seed certification
India	NBFGR, at Lucknow, Uttar Pradesh	Recognized nodal agency for formulating legislation on aquatic animal health certification and quarantine
Cambodia	–	–
China	National Certification Committee on Aquatic Wild and Bred Varieties (NCCA-WBV)	Certification of genetically improved varieties
Indonesia	Director General of Aquaculture	Certifies the hatchery management and production process, food safety, and traceability
Pakistan	–	–
Philippines	BFAR	Certifies and distributes improved tilapia strains
Thailand	Department of Fisheries	Guidelines and codes developed on aquaculture farm standardization
Sri Lanka	–	–

Fisheries Policy of 1998). Policy related to the freshwater fish seed industry:

- Leasing of government tanks, ponds, and other similar water bodies to targeted poor or unemployed youth, both men and women, for fisheries as means of their livelihood;
- Undertaking integrated aquaculture in inundated rice fields;
- Providing support to the establishment of hatcheries and nurseries in both public and private sectors to produce the required fingerlings for stocking in open waters and for aquaculture as well as for the establishment of the fish seed industry;
- Encouraging women in aquaculture;
- Developing guidelines for proper application of lime and fertilizer based on pond status.

On location-specific assessment of soil quality, the new Fisheries Law (draft) of Cambodia describes aquaculture management comprehensively (DoF, 2004a as quoted by So et al., 2005). The following inland aquaculture operations require permission from the Department of Fisheries:

- A pond or a combination of ponds with a total area larger than $5000 \, m^2$;
- A pen or a combination of pens with a total area larger than $2000 \, m^2$;
- A cage or a combination of cages with a total area larger than $15 \, m^2$.

However, it is not clear whether this regulation is aimed at minimizing negative impacts on the environment. In order to minimize negative environmental impact, the cumulative impact of total aquaculture establishments in an area should be taken into consideration rather than the total area of a single establishment. However, most of the small-scale fry nursing in both ponds and hapas and household aquatic production systems will not come under scrutiny under the new law due to the small area. In China, the amended Fisheries Law 2000 has more focus on fish seed production by including aquatic seed management. Under this law, any new aquaculture species can be propagated subject to certification by the NCCA and approved by the Ministry of Agriculture. Indonesia has several ministerial decrees issued by the Ministry of Agriculture concerning the fish seed industry. These regulations are concerned with: (1) managing the supply and distribution of fish seed, (2) prioritizing the production of quality domestic seed in required quantities, and (3) implementing technical standards of breeding to guarantee the quality of seed produced. Governments have also recognized aquaculture within the context of rural development and poverty alleviation policies (e.g., Vietnam). Government policy in all Mekong riparian countries has been supportive of aquaculture and some governments have production and earning targets for future development. These pro-aquaculture policies have supported investments in research, infrastructure, education, and extension that have contributed significantly to the growth of aquaculture in the past 10 years (e.g., Thailand, Vietnam) and as a result there has probably been less attention paid to issues of inland fisheries (Phillips, 2002).

Microcredit is crucial to ensure the participation of small-holders and resource-limited farmers. Credit systems in rural areas are undeveloped and difficult to access. Nevertheless, relatively well-developed microfinance systems exist in Thailand and Vietnam. The Vietnam Bank for Agriculture and Rural Development, e.g., provides loans for freshwater fish culture activities provided farmers have a "red book" demonstrating "ownership" of the land as collateral. Such microcredit systems will not benefit resource-poor farmers lacking collateral. As an alternative to fish seed certification, voluntary codes and guidelines are implemented in Thailand from hatchery management and operations to outgrowers and processors The Department of Fisheries certifies Code of Conduct products from hatcheries to the processor to enhance customer/consumer confidence on quality, safety, and environmental friendliness. The guidelines on Good Aquaculture Practices help the hatcheries and farms focus on siting of farms/hatcheries, farm/hatchery management, feed, chemical input, animal health, farm/hatchery sanitation, postharvest, and data collection. The Department of Fisheries has a monitoring mechanism to carryout regular monitoring for hygiene and good aquaculture practices of hatcheries/farms and targeted

30,000 hatcheries and farms to be certified for Code of Conduct and Good Aquaculture Practices by the year 2004. This monitoring is, however, emphasized on the use of feed and therapeutants. In a situation where large quantities of fish seed are imported and distributed throughout the country, a fish seed certification and quarantine process to preclude transboundary movement of aquatic animal diseases is necessary. This concern is growing in Cambodia where there is no certification or quarantine process, but is included in the draft new Law on Fisheries. To start with, a fish seed quality certification process may be initiated for the fish seed importers to observe.

8.4.3 Seed Certification as Proposed Some Eastern States of India

8.4.3.1 Act to regulate the quality of fish seed for production, marketing, and stocking of water bodies in State of Assam, India

Aquaculture development, in all sectors, is primarily dependent on the steady supply of quality seed to the culture sector, which raises a significant question for sustainability due to the dire paucity of quality seed both from natural and artificial sources. The natural source is declining rapidly and at the initial stage this gap in the supply chain was substituted by the successive development and dissemination of induced breeding technology to the field. Many parts of the world have ready availability of seed to farmers. Although seed of key cultured species (e.g., cyprinids, catfish, tilapia, shrimp, and prawn) are now produced in large quantities, poor quality is increasingly seen as a major impediment to the success of aquaculture. However, it must be remembered that in all aquaculture systems, stocking quality seed does not necessarily ensure a successful crop. The success depends on a number of other factors including poststocking management practices (e.g., water quality, feed, and disease management). There is a need to very clearly understand the relationship between quality seed and poststocking success on a species-by-species level for important cultured species. Otherwise there is a danger of all crop failures (e.g., poor growth, mortality, low survival) being attributed to seed and seed quality alone. Such a misconstrued approach might limit the opportunities to look for poststocking management interventions that could contribute to a successful crop.

The Act known as the Assam Fish Seed Act, 2005, clearly defines each step, principle, and procedure. The hierarchy includes:

Certification Officer under section 14;

State Level Seed Committee under section 3.

The Act presents definitions on crucial words, terms, and sentences so that there is no confusion between fish farmers and policy makers. For example, "Breeding," which includes "Control Breeding," "Cross Breeding," and "Reciprocal Breeding," "Test Cross & Back Cross" are clearly defined. Other clear explanations of the different terminology include:

1. Fish egg released by brood fish. Spawn length up to 8 mm;
2. Fry length above 8–40 mm;
3. Fingerling length above 40–80 mm;
4. Advance fingerling above 80–150 mm;
5. "Fish seed producer" means a fish farmer who produces fish spawn by induced breeding and controlled breeding and also by using a "hapa," which is a cloth or nylon enclosure.

In the process of certification "Fish Seed Grower," and "Fish Seed Importer," and "Fish Seed Exporter" are three steps, meaning a fish seed seller who sends fish seed in live condition outside the State. "Hatchery" means an integrated infrastructure for fish breeding, incubation, hatching, and marketing of fish seed.

"Induced Breeding" means the technology of fish breeding where mature male and female fish are induced to breed through administration of hormones; "Prescribed" means prescribed by rules made under this Act; "State Government" means the Government of Assam; "Water Bodies" means the ponds, tanks, beels, streams, rivers, derelict water reserves, swamps, and other water reserves where fish populations exist.

8.4.3.1.1 Constitution of a State Level Seed Committee

The State Government shall, by notification, constitute a State Level Seed Committee (SLSC) to exercise the powers conferred on and the functions assigned to it under this Act. The Act also provides guidelines for the production, marketing, sale, storing, supply, export, or import of fish seed of the species as provided in schedule I without first being registered with the District Fishery Development Officer (DFDO) or any other officer authorized in this behalf by the State Govt. under section 7 and obtaining a licence therefore. The form for applications for registration and licence shall be such as may be prescribed.

8.4.3.1.2 Procedure for Registration and Issuing Licence

1. Any person, society, association, farm, or body of persons, desirous of carrying out production, marketing, sale, storing, export, or import of fish seed shall apply before the 31st day of January every year in the prescribed form before the District Fishery Development Officer or any other officers authorized by the State Govt. in this behalf for registration and obtaining a licence as a fish seed grower, producer, seller, stockist, exporter, or importer as the case may be.

2. On receipt of the application, the District Fishery Development Officer or any other officer authorized in this behalf, shall make such inquiry as to the infrastructure available with the applicant, the purpose, the financial capability, and other relevant factors and thereafter register and issue a licence to the applicant on or before 28th February every year on payment of such fees as may be prescribed or refuse to register and issue a licence after recording reasons in writing, if he is not satisfied with the infrastructure available with the applicant or financial capability or other relevant factors of the application required for the purpose which shall be communicated to the applicant.

3. Any applicant aggrieved by the decision of the District Fishery Development Officer refusing to register and issue a licence may within 15 days from the date of communication of such order shall prefer and appeal before the Zonal Deputy Director of Fisheries who shall dispose of the appeal within 15 days from the date of receipt and his decision thereon shall be final.

4. Any person, society, association, fund, etc. to whom a licence has been issued shall be called to attend a training course to be organized by the Fishery Department within the 30th day of April every year; failing to undergo and complete the said training as and when called, the registration and licence granted to the applicant shall be liable to be cancelled; provided that those persons, societies, farms, etc. who have already undergone the said training need not be required to attend this training course.

5. In case of a society, association, farm, or body of persons who have been called for training course may depute any of their office bearer or other authorized representatives to undergo the training course.

6. All existing fish seed producers, growers, sellers, stockists, exporters, importers, etc. shall apply for registration and licence under this section to the District Fishery Development Officer within 3 months from the date of commencement of this Act. The District Fishery Development Officer may register and grant a licence to the applicant provided he is satisfied as to the criteria as laid down under subsection, failing to apply and obtain registration and licence under this subsection shall entail the existing fish seed producer, grower, seller, stockist, exporter, importer, etc. to discontinue the production, growing, selling, stocking, exporting, or importing of fish seed forthwith.

8.4.3.2 Renewal of Registration and Licence

1. The registration made and licence issued under section 7 shall be valid for a period of 1 year from the date of such registration and issue of the Licence.

2. The registration and licence may be renewed by the District Fishery Development Officer for a further period of 1 year at a time on receipt of application for renewal and on payment of such fees as may be prescribed and if the requisite provisions as laid down in this Act are fulfilled. The District Fishery Development Officer may refuse to renew the registration and licence, for reasons to be recorded in writing, if he is satisfied that the fish seed producer has fail to comply with the provisions of the Act and the conditions as laid down in the licence, giving him a reasonable opportunity of being heard.

Any fish seed producer aggrieved by the order of the District Fishery Development Officer may, within 15 days from the date of such refusal, preferred and appeal to the Zonal Deputy Director of Fisheries who shall dispose of the appeal within 15 days from the date of filling and which decision thereon shall be final.

8.4.3.2.1 Cancellation of Registration and Revocation of Licence

The District Fishery Development Officer or any other officer authorized by the State Govt. under section 7, may cancel the registration and revoke the licence granted to any person, society, association, farm, or body of persons if the registration and licence has been obtained by the holder thereof by misrepresentation of facts and has failed to comply with the conditions subject to which the registration and licence have been issued or has contravened the provisions of this Act and the rules framed thereunder, which are required to be compiled with by the holder of the registration and licence, by giving him a reasonable opportunity of being heard. Any person, society, association, farm, or body of persons aggrieved by the cancellation of registration and revocation of licence may, within 15 days from the date of communication of such order, preferred and appeal to the Zonal Deputy Director of Fisheries who shall dispose of the application within 15 days of failing of the appeal and his decision thereon shall be final.

8.4.3.2.2 Establishment of Brood Fish Bank and Seed Bank and Maintenance of Stock

1. The State government shall establish at least one Brood Fish Bank and Seed Bank in each fishery zone in the government fish and fish seed production farms available within the zone for rearing of breeders of standard endemic species exclusively from wild collected spawn and to supply quality fish seed to fish seed producers for raising quality brood fish for subsequent production of quality fish seed and shall notify the quality of brood fish and seed from time to time for compliance by seed producers.

2. Crossbreeding design of different hatchery stocks of each exotic species shall be prepared and practiced in the Brood Fish Bank for maintenance of the genetic diversity of exotic and improved endemic species. The Brood Fish Bank shall produce fingerlings from a large number of breeders at a time to ensure good genetic diversity, healthy breeders, and quality seed. No person, association, farm, or body of persons shall collect spawn from any natural source except by the Brood Fish Bank and Seed Bank established under subsection (1); Provided that District Fishery Development Officer may accord permission for collection of spawn from natural sources to any university or Research Institute for the purpose of scientific research, etc. and the said permission shall be limited to that purpose only. The Brood Fish Bank and Seed Bank may either distribute to other hatcheries or after initial rearing distribute either as fry or fingerlings. The wild origin of the spawn shall be certified by the concerned District Fishery Development Officer.

8.4.3.2.3 Maintenance of Brood Fish Bank and Seed Bank

1. In order to maintain quality production of brood fish, the Brood Fish Bank and Seed Bank as well as the fish seed producers shall maintain the following standards, namely:

   a. For production of brood fish during breeding operations exotic and wild or improved endemic species, equal numbers of males and females should be used for reproduction so as to facilitate one male fertilizing eggs from one female and the male once used shall not be used again to fertilize eggs of another female during the same breeding season.

   b. Brood fish candidates of exotic species should be produced only in the Brood Fish Bank and Seed Bank established under section (9), the spawn produced for brood stock rearing should be originated from a minimum of 50 numbers of parents (25+ 25) to avoid inbreeding.

   c. In no circumstance is anyone other than the Brood Fish Bank and Seed Bank allowed to produce brood fish in order to avoid negative selection, inbreeding depression, genetic drift, and genetic introgression of reproduced species.

   d. Origin and age of brood stock shall be recorded and maintained in a stock register in each hatchery.

   e. The Brood Fish Bank and Seed Bank, as well as the fish seed producers, shall use the spawn measuring cup as standardized by the Indian Council of Agricultural Research (ICAR).

   f. Brood fish weight shall not be less than 3 kg in the case of catla, silver carp, and grass carp; 1.5 kg in the case rohu and mrigal (mirka); 1 kg in the case of common carp, and 0.5 kg in the case of kuri.

g. At least 30% of brood fish stock shall be replaced every year by the Brood Fish Bank and Seed Bank as well as by the fish seed producers.

h. Propagation of exotic fish species, such as *Clarias gariepinus* (Thailand magur) *Orocochromis* sp. (tilapia), *Aristichthys nobilis* (bighead carp) are completely banned for breeding for the time being; Provided that the State government may, from time to time, specify, by notification in the Official Gazette, the name of species breeding of which shall be banned for the interest of protection of the endemic species, as may be deemed necessary.

8.4.3.2.4 Species of fish to be covered under this Act

On the date of commencement of this Act the species of fish as enumerative in the schedule-I shall be come under the purview of this Act. The State government may, by notification in the official gazette, amend, vary, alter, add to, modify, or omit any species of fish from schedule-I and thereupon said schedule shall be deemed to have been amended, varied, altered, or modified accordingly.

8.4.3.2.5 Regulation on Seed Sale

The State government may issue guidelines, in consultation with the committee from time to time and in conformity with the provisions of this Act, on production, growing, sale, stock, export, import, or other matters as may be considered necessary and the said guidelines shall be binding on all the fish seed growers, producers, sellers, stockists, exporters, or importers, as the case may be. The fish seed importer shall take all necessary steps to treat the fish seed with disinfectants in accordance with the norms laid down from time to time by the Indian Council of Agricultural Research to avoid spread of diseases, if any, such as quarantine measures, prior to marketing. No person shall, himself or by any other person on his behalf, carry on the business of selling, keeping for sale, offering to sale, bartering, or otherwise supplying any fish seed of any variety unless

1. Such fish seed are produced or raised by registered producers or growers;
2. Such fish seed is strong, healthy, and disease-free;
3. Such fish seed is transported in well-aerated containers without overstocking in packages; and
4. Such fish seed is duly certified by the Certification Officer empowered under section 14.
5. The fish seed producer and seller use the spawn measuring cup as standardized by the Indian Council of Agricultural Research.

8.4.3.2.6 Fish Seed Certification

Every fish seed producer, fish seed grower, and fish seed importer, shall apply in the prescribed form and obtain a Fish Seed Certificate from the officer mentioned under subsection in the form as may be prescribed, before sale, marketing, transport, and export of fish seed. The District Fishery Development Officer or any other officers authorized by the State government shall be authorized officers for certification of fish seed. The District Fishery Development Officer or any other officer authorized by the State government under subsection (2) before issuing a Fish Seed Certificate shall make an inspection and conduct such other inquiry as may be considered necessary as to the health of the fish seed taking into consideration the desired factors as provided in Schedule II. The Fish Seed Certificate shall be in such form, as may be prescribed. The District Fishery Development Officer or any other officers authorized under subsection (2) may refuse to issue a Fish Seed Certificate if the condition of health of the fish seed does not conform to the desired factors as provided in Schedule, for reasons to be recorded in writing.

8.4.3.2.7 Fish Seed Regulation

The District Fishery Development Officer or any other officers authorized by the State government in this behalf shall be the Fish Seed Regulator for the purposes of this Act and shall exercise the following functions, namely:
1. To inspect and take samples of fish seed of any variety from any person producing, storing, or selling such seed to a purchaser or any other person.
2. To exercise such functions in respect of entry, search, and seizure as has been entrusted to him as such officers in section 16.
3. To examine any record, register, document, or any other materials, objects found in any place, and seize the same if he has reason to believe that it may furnish evidence of the commission of an offence.
4. To furnish a monthly report to the respective Zonal Deputy Director of Fisheries in respect of such matters as may be prescribed.
5. To exercise such other powers as may be necessary for carrying out the purposes of this Act and rules made thereunder.

8.4.3.2.8 Power of Entry, Search, Arrest, and Detention

Notwithstanding anything contained in any other law for the time being in force, any officer of the Fishery Department not below the rank of Fishery Extension Officer, if necessary with assistance of police or Civil Administration or any Police Officer, not below the rank of a Subinspector in the absence of a regular Subinspector may, or any Assistant Subinspector in the absence of a regular Subinspector may, if he has reasonable grounds for believing that any person, association, farm, or group of persons have committed an offence under this Act: require any such person, etc., to produce for inspection any fish seed, brood fish, container, article, appliance, tool, vehicle, boat, steamer, vessel, carrier, and any other materials in his possession and books, licence, and records which are required to be maintained by him under the provisions of this Act; enter at all reasonable times any place, premises, vehicles, vessel, carrier, etc. used for suspected to be used for production, storage, sale, export, import, or transportation etc. of fish seed for inspection or otherwise; stop any vehicle, vessel, boat, steamer, or any other carrier for the purpose of conducting search of enquiry; seize any fish seed, brood fish container, vehicle, vessel, boat, steamer, carrier article, appliance, tools, or any other things which are used for committing any offences under this Act and unless he is satisfied that such person or persons responsible for commission of such offence shall appear and answer any charge which may be preferred against him, arrest him without warrant; prepare a list of seizure of the fish seed, brood fish, articles, vehicles, vessels, or any other things seized under clause in presence of at least two witnesses of the locality which shall also be signed by the person from whom the articles, things, vessels, vehicles, fish seed, brood fish, etc. have been seized and by the officer concerned and deliver a copy of the seizure list so prepared to the person concerned; may release the seized materials and things on the execution by the owner thereof of a Bond for the production of the seized article so released, if and when so required, before the magistrate having jurisdiction to try the offence on account of which the seizure has been made; provided that the stock of fish seed or brood fish so seized shall not be disposed of by the owner thereof within 30 days from the date of seizure and after the expiry of 30 days the said stock may be disposed of by the owner thereof with due permission from the Magistrate having jurisdiction to try the offence, showing the reasonable grounds to the satisfaction of the Magistrate for its disposal; release the person arrested under clause on the execution by him of a Bond of any amount which shall not exceed 20,000 rupees with a surety of the like amount for ensuring his appearance and answering the charge preferred against him; forward a copy of each of the seizure lists, release of seize articles, arrest, and release order of the person concerned along with a copy of the bond and surety obtained in respect of release of the arrested person to the Magistrate having jurisdiction to try the offence on account of which the seizure has been made.

8.4.3.2.9 Offence to be Cognizable and Bailable

The offences under this Act shall be cognizable and bailable within the meaning of the Code of Criminal procedure, 1973 (Act 11 of 1974).

8.4.3.2.10 Cognizance of Offence

No court other than Judicial Magistrate of Offences the First Class shall take cognizance of any offence under this Act except on the complaint made by any officer of the Fishery Department, not below the rank of a Fishery Extension Officer or a Police Officer not below the rank of Subinspector.

8.4.3.2.11 Offence to be Tried Summarily

The offence committed under this Act shall be tried summarily under the provisions of the Code of Criminal Procedure, 1973 (Act II of 1974).

8.4.3.2.12 Penalty

If any person contravenes any provision of this Act or rules made there under or prevents any officer from exercising any power conferred on him under this Act, he shall be punished with a fine which may extend to 3000 Rupees, if the offence is committed for the first time and in the event of commission of the second and subsequent offences under this Act, he shall be punished with imprisonment of either description for a term which may extend to 6 months or fine up to 5000 Rupees or both.

8.4.3.2.13 Protection of Action taken in Good Faith

No suit, prosecution, or other legal proceeding shall lie against the State government or any officer or other employee of the State government for anything which is done in good faith or intended to be done under this Act.

8.4.3.2.14 Officers to be Public Servants

Every officer exercising any of the powers conferred by this Act shall be deemed to be public servants within the meaning of section 21 of Indian Penal Code (Central Act 45 of 1860).

8.4.3.2.15 Declaration of Minisanctuary

Within 3 months from the date of commencement of this Act the State government shall, by notification in the official Gazette, declare such number of close water bodies, whereupon the ownership rights belong to the State government, as minisanctuaries containing an area of at least $5000\,m^2$ for each 20 hectare 10 water surface under culture-based management to ensure the survival of a breeding population of wild varieties or nonstock species of fish wherein harvesting may be done every second year with the exception of hook and line fishing with live fish bait to prevent damage to small fish by predatory fish. Before declaring any minisanctuary under subsection (1), the State government shall take into consideration the geographical, ecological, geomorphologic, floral, faunal, or zoological associations, importance, or requirements as may be necessary for declaring the said water bodies as minihatcheries.

8.4.3.2.16 Removal of Difficulties

If any difficulty arises in giving affect to any of the provisions of this Act the State government may, by order not inconsistent with the provisions of this Act, remove the difficulties.

8.4.3.2.17 Power of the State Government to Make Rules

The State government may by notification in the official gazette make rules for carrying out the purposes of this Act. In particular and without prejudice to the generality of the foregoing power subrules may provide for all or any of the following matters, namely:

1. Procedure for transaction of business by the committee at its meetings under section 3(3);
2. Format for submission of annual report by the committee under section 5(e);
3. Honorarium and allowances and other conditions of service of the nominated members of the committee under section 4(2);
4. Form for application for registration and licence under section 6(2);
5. Fees for registration and licence under 8(2);
6. Form for application for Fish Seed Certification and form for Fish Seed under section 14(1);
7. Matters on which monthly report to be furnished by Fish Seed Regulator under Section 15(iv) 3. All rules made by the State government under this Act shall, as soon as may be after they are made, be laid before the Assam Legislative Assembly while it is in session, for a total period of not less than 14 days which may be comprised in one session or two or more session, and shall, unless some later date is appointed, take effect from the date of their publication in the official gazette subject to such modification or annulments as the Assam Legislative Assembly may, during the said period agree to make, so however, that any such

modification or annulment shall be without prejudice to the validity of anything previously done there under Schedule-I (See Section 6(1)).

Scientific Name of Indigenous Species and Local Name

Scientific name	Local name
1. *Acrosssocheilus hexagonolepis or Neolissocheilus*	Boka pitha
2. *Anabus testudinesu*	Kawoi
3. *Catla catla*	Bhakua
4. *Channa marulius*	Sal
5. *Channa punctatus*	Goroi
6. *Channa striatus*	Sol
7. *Cirrhinus mrigala*	Mirika
8. *Chirrhinus rebba*	Lasshim
9. *Clarias batrachus*	Magur
10. *Heteropneustes fossilis*	Singhee
11. *Labeo bata*	Bhangon
12. *Labeo calbasu*	Mali
13. *Labeo dero*	Silghoria
14. *Labeo dyocheillus*	Bokaghoria
15. *Labeo gonius*	Kurhi
16. *Labeo nandina*	Nandani or nandoni
17. *Labeo rohita*	Rou
18. *Mystus aor or*	Aorichthys aor Bheu
19. *Mystus menoda*	Gagol
20. *Mystus seenghala or*	macrons seenghala Arii
21. *Notopterus chitala*	Chital
22. *Ompok bimaculatus*	Pabhoo
23. *Ompok pabho*	Pabho
24. *Puntius sarana*	Seni puthi
25. *Rita rita*	Ritha
26. *Tor putitora*	Jugnga pitha
27. *Tor tor*	Pithia
Exotic species	
1. *Ctenopharyngodon idella*	Grass carp
2. *Cyprinus carpio*	Common carp
3. *Hypohthalmichthys molirix*	Silver carp
4. *Puntius javanicus*	Java puthi

List of Fish to be Covered by "ASSAM FISH SEED ACT, 2004".

Schedule-II (see Section 14(3) Desired Factors for Selecting Healthy Fish Seed

Desired fish seed	Not desired
1. Vigorous and actively swimming	1. Lethargic and abnormal swimming behavior, dark and abnormal body
2. Normal body color, soft to touch	2. Abnormal body color, rough to touch, and excessive mucus secretion
3. No sign of any patches, spots, or lesions	3. Appearance of discolored patches, spots, cysts, lesions
4. Fins intake complete	4. Broken fin tips
5. Gill raker deep red and without	5. Discoloration/broken gill, lamellac, any sign of hemorrhage, spots, or cysts, appearance of hemorrhage, spots cysts, etc.
6. Quick response to external stimuli	6. No or feeble response to external such as tapping, etc. stimuli

(Ref. M.K. Deka, Commissioner & Secretary, Govt. of Assam Legislative Department.)

"Container" means leak proof polythene of 18 in. Diameter containing well-aerated oxygenated water duly sealed and packed in a tin, carton, sack, or bag, etc.

8.4.4 Recommendations of the Inland (tank and ponds, river and reservoir) / Fisheries sub-committee

8.4. 4.1 Issues on Quality Fish Seed

8.4.4.1.1 Hatchery Accreditation and Seed Certification

State to formulate/implement a fish seed certification norms based on the guidelines likely to be issued by Govt. of India.

1. Fish seed producers/hatchery owners in the state are required to be duly registered in the event of following conditions
   - Production capacity beyond 1 million spawn/year
   - Fry rearing capacity of more than 1 million spawn
   - Coming under the ambit of other regulatory rules mentioned in the form of registration and should bound by the legal provisions to discourage defaulters or violators of hatchery breeding code of conduct.

2. District level fisheries officials will be designated authorized officers for enforcing registration
3. Training should be arranged by the state for the hatchery managers in quality seed production by the state (capacity over 10 million)
4. Fish seed produced from the registered units will enjoy the right to free movement within the state
5. Inter state movement of fish seed should necessarily require additional clearance from the authority based on the application to that effect by the registered hatcheries only
6. Director of Fisheries will be the appealate authority or of a similar status as decided by the Govt. of Orissa for dispute settlement, if any
7. National Research Laboratories will act as referral bodies for applet authority seeking clarifications, if any.
8. Registered units should specify the following information, renewable biannually
   - Source of brood stock
   - Quality of stock
   - Production capacity
   - Provision of stock exchange
   - Equipped with skilled/trained manpower
   - Provision of fish health monitoring
   - Declaration on avoidance of any un-ethical hybridization
   - Provision of good water supply
   - Scope for safe disposal of farm/hatchery wastewater
9. **a.** Seed Transport
   - Exemption on octrai and sales tax
   - Power of arbitration vested on National institutes (Farmer vs. State, State vs. State)

   b. Fish Feed
   - All fish feed mills of the state to be registered with the Department of Fisheries of any other body authorised by the state
   - Date of manufacture and expiry should be printed on the label
   - Composition of the feed should be cleanly specified on the label
   - Feed should not contain any banned drugs (antibiotics/steroids)
   - The feed should be exempted from the sales tax for enabling the farmer increasing the fish production
   - Quality of the ingredients used should be certified by recognized lab.

   c. Non-conventional Aquaculture
   - Water logged areas to be developed for fish production through aquaculture

- Paddy fields where controlled irrigation is possible should be stocked with fish/prawn seed
- Burrow pits alongside Highway/Rly tracks to be brought under fold of pisciculture
- Integration of pisciculture with location specific agricultural/horticultural/live stock/forestry/selviculture/sericulture activities
- Utilization of domestic sewage for fish production should be encouraged wherever possible following the technology by CIFA, so that the affluent usually discharged to river and other water ways should be used for increasing fertility of different waterways

d. Reservoir Fisheries

- All open water bodies MIP/Kata/Irr tank/Percolation tanks/Reservoirs to be brought under the fold of culture based capture fisheries
- Stocking of reservoirs should be made as per the recommendations of IIM/CIFRI
- Total ban on stocking of exotic carps
- Strict conservation measures to be followed as per the norms of the Biodiversity acts
- Observance of regulation on fishing, including net and mesh size or less, as per OSRFP 2004 (.................)
- Mandatory provision of one hatchery of required capacity adjacent to a reservoir or for a few adjacent reservoirs
- Provision on basic infrastructure *viz.*, ice/boat/net/road/captive NT/cages/fish landing shed etc.
- Provision of Ice plant, cold storage, etc. and transport facility should be promoted by the state
- State should organize marketing channel and establish linkages

e. Development of Riverine Fishery

- Breeding and seed production of riverine species and river ranching by the state
- Observance of closed season (15 June – 15 August) and development of alternative means for livelihood of fishermen
- Control measures for riverine pollution (domestic / industrial) with special reference to aquatic biodiversity
- Shallow pools adjacent to rivers to be protected as sanctuary
- Total ban of an introduction/transplantation of exotic species (dynamite, explosive, poisons, zero mesh nets)
- Restriction of fishing by destructive methods
- Proper policy for maintaining minimum environmental flow in rivers

- Need for collaborative efforts through stake holders to improve and protect the habitat which otherwise adversely affect fishery and biodiversity wherever possible

f. Health and Disease Monitoring of Fish and Shellfish
  - Establishment of disease diagnosis, control and monitoring centres at district level
  - Promotion of Aqua labs / Aqua clubs / Aqua shops through adequate support system
  - Monitoring of key events of epidemic diseases by Headquarters and measures trails control in collaboration of National Institute as referral lab

g. Observance of responsible code of conduct as per the guidelines of FAO & Govt. of India

h. Environmental impact assessment studies to be conducted before any mega project cleared by the state (individual / regional basis)

i. Establishment of Database
  - Building accurate/dependable up-to-date database
  - Enhancing e-connectivity to the District/Block level to check and monitor the programme
  - Computerization of District level offices for efficient record keeping and updating of information
  - Accession and dissemination of information from the headquarter level to National Institute / organization with internet

j. Human Resource Development
  - **i.** Updating the technical skill and of the FEO's through information/developments of the state to all the FEO's of the state through
  - Refresher course / practical training
  - Awareness programme
  - Field visits including interstate exposure
  - **ii.** All the district level functionaries to be exposed to most of the relevant developments in other advanced states, phase wise
  - **iii.** Headquarter level/zonal level officials are to be exposed to the HRD programme (National/International) related to policy formulation and prioritization, project formulation of the R&D activities including management

k. Marketing
  - Qualitative post-harvest care of the product
  - Establishment of hygienic fish markets in urban areas
  - Eco-friendly bio-product disposal system
  - State level/regional mundies (wholesale markets) for rationale distribution
  - Establishment of district level market for fish and fish seed

- Creation of date base on fish marketing
- Provision of data-base for production/processing & export
l. Reassessment of resources in terms of cultivable pond/tank area and present level of productivity on GIS platform
m. Provision of adequate credit facility for small scale fish farming
n. Development of National consensus for insurance aquaculture sector

8.5 Fish Seed Marketing

The main actors involved in the fish seed marketing are hatchery and nursery operators, middlemen or seed traders, and fish farmers. The market channels for fish seed need to be relatively short and simple because of the high risk involved in selling the product due to its perishable nature. The number of actors involved in the fish seed marketing and thereby the length of the market chain is often related to the type of end product (spawn, fry, or fingerling) of the hatchery and the accessibility between the hatchery and the farmer. The simplest form is a hatchery operator selling directly to grow-out farmers, either through delivery or pick-up. This type of direct marketing from the hatchery to the outgrower is often seen when hatcheries produce fingerlings in their own nurseries and have established access to farmers. Some hatchery and nursery operators use agents to increase sales, especially in remote areas. Others sell directly and employ agents at the same time (e.g., Philippines), who resort to intermediaries to buy fry from other areas, which they nurse to fingerlings. Generally, agents obtain their incomes by a mark-up of price per fingerling or through preagreed commissions.

However, break-up of production cycle is inevitable on many occasions due to land pressure and the involved cost of accommodating nursery facilities in the hatchery premises. Break-up of production cycle into breeding and nursing offer opportunities for poor farmers to enter into fish seed production. As seen in Bangladesh fingerling marketing, it is generally through middlemen since very few fish farmers buy directly from nursery farms. About 80% of fingerlings are supplied to farmers through middlemen and some 20% of fingerlings are directly collected by farmers from nursery operators. In addition to this, when spawn are reared to fingerlings at the hatchery-cum-nursery, fingerlings are also reached to farmers through the same channel. Buy-back arrangements are made between the hatchery and nursery, hatchery to provide spawn or fry on credit at a given price to the nursery buy back fry or fingerlings from them at prefixed prices allowing adequate margin to the nursery operator (India). Sometimes NGOs and GOs are also seen in the market chain as buyers. NGOs buy fingerlings for the farmers who are in their development programs, while GOs buy indigenous varieties for the purpose of stock enhancement (e.g., Cambodia). Governments and public lending organizations are encouraged to generate income. As a result some government hatcheries focus on selling fingerlings directly to farmers, creating a competitive environment between public hatcheries and private hatcheries and nurseries. A competitive environment can make producers quality-conscious of their product.

The regional focus has shifted from centralized to decentralized seed production. Fast-growing and more easily bred fish species such as common carp, silver barb, and tilapia can be bred easily by farmers without the need for hatchery facilities, using only hapas submerged in a water body or in flooded rice fields. The small investment enables poor farmers to adopt the technology. Local production and seed trading networks can reduce the need for long journeys to transport fish, and thereby reduce transport cost and improve seed quality. However, potential constraints exist when the fish seed supply is dependent on small, isolated fish brood stock populations. Unless decentralized fish seed production includes appropriate breeding strategies to maintain the genetic quality of brood stock, the performance of the production stocks will decline. Appropriate interventions to improve management practices and regular replenishment of high-quality seed for bloodstock require concerted efforts through participatory approaches with farmers, government agencies, and NGO stakeholders to develop and institutionalize improved rural seed supply.

8.6 Economics of Seed Production

The major inputs involving nursery operations are mainly dependent on organic and inorganic fertilization prior to stocking with poststocking supplementary feeding, also considered an organic fertilization, as it is composed of mustard oil cake and rice bran (1:1 by wt.). The market price of fish seed varies on species type, strain, size, supply and demand deficit, quality, breeding season (prespawning, postspawning, and spawning), farmer preference, source, and mode of transportation (proximity to supply). As indicated in Chapter 2, Reproductive Cycle, Maturation, and Spawning, carp spawn and fry produced during prespawning and postspawning seasons (mainly by the large hatcheries) fetch a higher price than fry produced during the breeding season. However, normal practice among farmers is to purchase fry from the postspawning season at a low price and then use these fry for overwintering. The price of overwintered fry is

Table 8.6 Price Chart of Hatchlings or Spawn

Species	Scientific Name	Price (Taka/kg)
Rui	*Labeo rohit*	2000
Catla	*Catla catla*	2500
Mrigal	*Cirrhina cirrhosis*	1500
Silver carp	*Hypophthalmichthys molitrix*	1500
Grass carp	*Ctenopharyngodon idella*	2300–2500
Mirror carp	*Cyprinus carpio* var. *specularis*	2000–2300
Common carp	*Cyprinus carpio* var. *flavipinnis*	1800–2200
Tilapia	*Oreochromis mossambicus*	1600
Thai punti	*Puntius gonionotus*	1500–1800
Black carp	*Mylopharyngodon piceus*	25,000–30,000
Calbaus	*Labeo calbasu*	2000–2500
Pangus	*Pangasius hypophthalmus*	2500–2800

quite high as they grow faster compare to prespawning fry (Table 8.6). Among Indian major carp, catla seed is more expensive than rohu and mrigal. Wild-caught fry are also more preferred than hatchery-produced fry in terms of quality. Fingerlings produced by private hatcheries are often 20–30% more expensive than those from government hatcheries due to their higher quality. In Cambodia, this quality is measured in terms of distance involved in the transportation of fingerlings to the farmer; a short distance is preferred. In the Philippines, seed prices from government hatcheries have remained relatively stable; any price change is based on cost recovery. Genetically improved varieties such as SRT, GET, and GST are naturally more expensive than mixed tilapia fry/fingerlings. Nevertheless, a higher price does not always mean higher demand or preference. In Vietnam, even though there is improved an common carp strain, which has a 50% increased growth rate, higher survival rate, and attractive appearance over the local strain, upland mountainous ethnic Thai farmers prefer the local strain (Tien, 1993). This preference is drive by other factors beneficial to poor farming households such as: (1) ability to breed at the household level, (2) cheaper seed price with ready availability in close proximity, (3) better adaptability to shallow fields and terraced rice fields with flow-through water, and (4) disease resistance. The local strain in upland mountain areas appears to be the better strain and this is also reflected in the 10% higher market

Table 8.7 Average Price of Fry of Different Species

Species	Scientific Name	Price
Rui	*Labeo rohit*	1800–1900 (Tk/kg)
Catla	*Catla catla*	2200–2400 (Tk/kg)
Mrigal	*Cirrhina cirrhosis*	1400–1500 (Tk/kg)
Silver carp	*Hypophthalmichthys molitrix*	1300–1400 (Tk/kg)
Grass carp	*Ctenopharyngodon idella*	2200–2500 (Tk/kg)
Mirror carp	*Cyprinus carpio* var. *specularis*	2100–2200 (Tk/kg)
Common carp	*Cyprinus carpio* var. *flavipinnis*	2300–2400 (Tk/kg)
Thai punti	*Puntius gonionotus*	1400–1700 (Tk/kg)
Black carp	*Mylopharyngodon piceus*	24,000–28,000 Tk/piece
African magur	*Clarias lazera*	3–4 Tk/piece
Pangus	*Pangasius hypophthalmus*	2600–3000 Tk/piece

TK, Taka (Rupees), 1 TK = 69 US$

price (Edwards et al., 2000). In Cambodia, the price of strictly carnivorous fish fingerlings such as hybrid catfish and snakehead is 30% more expensive during the off-season (Table 8.7).

In Bangladesh, traders who purchase wild collected freshwater prawn larvae earn a margin of around Tk 500–700 per 1000 postlarvae (PL) (Dr. Nesar Ahmed, personal communication). The price of the hatchery-bred freshwater prawn postlarvae ranges from Tk 1200–2500 per 1000 PL and early seasoned PL receive a higher price than the full breeding-seasoned PL. It is hard to generalize the demand and supply status of fish seed in the region. Seed may not be currently a constraint at the national level in many countries in the region, but there may be problems of supply locally and at particular times of the year. Demand and supply status depends on species, season, country, and sometimes a region within a country. Often carp spawn supply exceeds the demand and forces small hatchery operators to change species. Small hatchery operators change species from carp to indigenous catfish, Thai pangus, and Thai koi in Bangladesh to avoid economic losses, while large hatchery operators scale down their production or dispose of the production early to keep the operating costs down. Selling of fry/fingerlings early will lead to low quality and poor productivity. Presently, there is a wide deficit between supply and demand of fish seed, particularly in Sindh, NWFP, and Balochistan provinces of Pakistan. Seasonality in demand is not observed in cases where hatchery and fry nursery operators cater to the needs of stock enhancement programs (Sri Lanka). Freshwater farming, including hatchery and nursery operators, provides opportunities for self-employment for operators and their families. Backyard/small-scale operators rely mainly on family labor. Large hatcheries and nurseries employ regular full-time workers and seasonal or casual labor for the purpose of pond preparation, stocking, and harvesting. There are times that there is an exchange of labor from members of the community and neighbors and the owner is expected to reciprocate the initiative by helping fellow farmers when needed. These situations arise in rural areas where there is an abundant supply of labor and limited employment opportunities, wherein men, women, and children participate in and assist hatchery and nursery operators. Fish breeding and fry nursing are the sole income sources for some households. Fish fry nursing at small-scale farmer hatcheries significantly contributed to the total family income. A survey conducted by Hasan and Ahmed (2002) revealed that fry nursing contributes 79.3% of the household income, while hatchery operation contributes 95.1%. Apart from aquaculture the hatchery and nursery operators earn money from other economic activities such as paddy cultivation, livestock-raising, and vegetable and fruit production. Costs and returns of freshwater fish farming vary substantially by production environments, type of technology, and species cultured and across countries. The same applies for fish seed production. The tabulated data in Table 8.8 are clearly related to specific countries, sites, and projects, and have a degree of variation which makes it difficult to derive generalized descriptions/conclusions. Nonetheless, there are fundamental relationships, e.g., as described by size, productivity, and factor costs, which can be considered, and at a culture system level it is possible to develop some trends and predictions on factors affecting future growth. In general, fish seed production is regarded as a profitable venture. Average profit margins of 290–300% have been reported (Nam & Leap, 2007) on fingerling production in farmer-operated and government hatcheries and an 800% profit margin for small-scale fry nursing in Cambodia. In contrast, the profit margin (95%) for fry nursing of carnivorous fish (*P. hypophthalmus*) is low as they do not have sufficient pond space, and brood stock maintenance of slow-maturing species is meant for government hatcheries. Maintenance of medium- (*L. hoevenii, O. melanopleura*) and fast-maturing (*B. gonionotus* and *B. altus*) brood stock may be within the reach of small-scale hatcheries as the estimated average maintenance cost is around 12 and 15 times less (US$50 and US$40 per year), respectively. The difference in returns reported for 500 m^2 for pond nursery rearing of carp in Sri Lanka is attributed to the management experience acquired

Table 8.8 Some Indicative Cost Profiles of Fish Seed Production Systems

(a) Capital Costs

Aquaculture System	Culture Practice	Capital Cost (US$)	Notes: Country Refers to Case Study in This Publication
Hatchery of 18 million hatchling production	Carp	100,000 Indian Rs	India/Ramachandra (1996)
Hatchery of 10 million hatchling production	Carp		
Pond nursery (500 m^2)	Carp (CBO managed)	707 US$/500 m^2	Sri Lanka
Pond nursery (3500 m^2)	Carp (CBO managed)	4950 US4/3500 m^2	Sri Lanka
Pond nursery (500 m^2)	Carp (privately owned)	752 US$/500 m^2	Sri Lanka
Hapa	*Macrobrachium* fry rearing	8.25/4–5 m^3	Bangladesh

(b) Total Operating Costs

Aquaculture System	Culture Practice	Capital Cost (US$)	Notes: Country Refers to Case Study in This Publication
Hatchery of 18 million hatchling production	Carp	103,125 Indian Rs	India/Ramachandra (1996)
Hatchery of 10 million hatchling production	Carp	42,100 Rs	India/Das and Sinha (1985)
Pond nursery (0.1 ha)	Carp	2000 per crop	India/Das and Sinha (1985)
Pond nursery (0.2 ha)	Carp	7100 per crop	India/Das and Sinha (1985)
Pond nursery (500m^2)	Carp (CBO managed)	598 US$/annum	Sri Lanka

Pond nursery (3500 m²)	Carp (CBO managed)		4184 US$/annum	Sri Lanka
Pond nursery (500 m²)	Carp (privately owned)		827 US$/annum	Sri Lanka
Hapa	Macrobrachium fry rearing		76.4/4–5 m³/4–6 weeks	Bangladesh

(c) Financial/Economic Performance

Culture System	Species	Production Cost (US$/kg)	Net Revenue (US$)	Profit Margin (%)	Return on Operating Investment (%)	Notes
Hatchery of 18 million hatchling production	Carp	5729 Rs/1million hatchlings	50,875		49.33	India/Ramachandra (1996)
Hatchery of 10 million hatchling production	Carp	4210/1 million hatchlings	20,400		48.4	India/Das and Sinha (1985)
Pond nursery (0.1 ha)	Carp		3400 per crop		170	India/Das and Sinha (1985)
Pond nursery (0.2)	Carp		1900 per crop		26.8	India/Das and Sinha (1985)
Pond nursery (500 m²)	Carp (CBO managed)		558 US$/annum	48	93	Sri Lanka
Pond nursery (3500 m²)	Carp (CBO managed)		3906	48	93	Sri Lanka
Pond nursery (500 m²)	Carp (privately owned)		1311 US$/annum	61	158	Sri Lanka
Hapa	Freshwater prawn fry rearing	0.05/juvenile	39.1	32	51	Bangladesh

by the private nursery operators over a longer period of involvement than that of Community-based Organizations (CBOs). However, the calculations of reported returns do not include certain fixed (loan repayment, lease/rent, etc.) and variable (labor and energy) costs. When these fixed and variable costs are included in the total operating costs the returns may be less attractive. The effect of labor on the returns may be significant and can be minimized by determining an optimal economic size. Labor is required for fertilization and feeding, and for guarding where ponds are distant from homesteads. Total labor requirements are likely to be in the region of 1–2 Full Time Equivalents (FTE) per ha, which is a fairly standard figure for pond culture throughout the region. In the case of small homestead nursery ponds or hapas installed in water bodies near homesteads, a significant part of this labor input can be provided by women and children, and opportunity costs may be low, implying higher rates of return and/or lower production costs. There are five important risk factors for nursery rearing:

- Loss of stock due to flooding;
- Loss of stock due to disease;
- Loss of stock due to poaching;
- Poor growth and survival due to poor water quality, especially toward the end of the dry season/production cycle;
- Inadequate monitoring due to lack of onsite management tools.

Nursing of postlarvae of freshwater prawns in hapa systems, even though profitable and generating healthier returns than fish fry rearing in ponds due to short cycle length (4–6 weeks) requires higher levels of investment and access to brackish water. However, the short cycle length (4–6 weeks) means that turnover is rapid, and any loan interest will be minimal. Indicative cost structures of alternative seed production systems need to be further analyzed to determine these activities fall out into three major groups: the labor-intensive systems, the feed-intensive systems, and the fertilizer-intensive systems. The first and third of these categories are of most interest from the perspective of poverty alleviation. Farmers at certain suitable located site, which receive fish seed from several regions in the country, transported on train (Jessore, Bogra, and Rajshahi), have a sales turnover of 100–125 million fingerlings every year (NEFP, 1993–1994). The fish seed distribution companies are active in northeast and northwest regions of China, where many small-scale farmers are concentrated, to facilitate distributions of fish seed received from the southern region of China. Such network arrangements involve fish seed collectors at local/village level who supply fish seed to regional collectors/distributers (Indonesia). The regional collectors/distributors provide funds for local collectors to make upfront payments or to provide loan payments for farmers at the village level to commit them to supply. Therefore network arrangements are often induced by the distance of delivery. The exotic fish seed are distributed by companies/agents who import fish seed from abroad, and are distributed through their sales network or outlets (eels in China, Cambodia).

Fish seed traders are the most critical actors in a complex network linking hatcheries and seed nurseries to fish farmers as they not only facilitate the seed supply, but also provide advice and disseminate knowledge on advanced farming to fish breeders and farmers. Seed trading is mostly a seasonal occupation that, in most places, is confined to a particular fish seed-rearing season. They use several modes and facilities to transport fish seed ranging from foot to motor vehicles and clay or aluminum pots to aerated containers or sealed polythene bags. While seed traders are useful in disseminating knowledge, they can also account for significant losses of seed during long journeys and reduce quality. Therefore, improvements to the modes of fish seed transportation are equally important to maintain fish seed quality.

This important role of fish seed traders has urged training them on improved handling of fish seed during transportation and basic fish culture to help dissemination of knowledge to the farmers (Northwest Fisheries Extension Project in Bangladesh). Key financial performance measures returns on labor and profit margin need to be further analyzed. Profit margin measures profitability in terms of the ratio between profit and gross revenue, and is an indicator of both susceptibility to falling price, and relative production costs. Return on labor is a measure of the total net income generated per unit of labor expended and represents the maximum average wage for people engaged in the enterprise. If hapa and small pond nursing score well against these financial parameters, the economic benefits are likely to be more widely spread.

8.7 Information and Knowledge Gaps

As technologies for small-scale aquaculture are now largely in vogue, an information sharing mechanism on captive breeding, as also fry/fingerling production on a commercial scale, will help build capacities in countries where these technologies are not well developed. Dissemination, training, and specific technology requirements vary regionally and countrywise. Identification of preferred species and the relevant technologies should be disseminated, up to the village level, to encourage backyard activities.

Chemicals in the form of pesticides, disinfectants, drugs, antibiotics, hormones, anesthetics, and probiotics are used in hatcheries as well as in nursery operations. They are often essential components in such routine activities as tank and pond preparation, water quality management, fish seed and brood stock transportation, feed formulation, manipulation and enhancement of reproduction, growth promotion, disease treatment, and general health management. Our study indicates, in the absence of aquaculture-specific chemicals, that fish breeders and hatchery owners often use chemicals of veterinary origin or as directed by the agents, distributors, wholesalers, and sales persons. Often, the fish breeders, hatchery owners, and ignorant farmers are befouled by these agents as there is an almost complete absence of any transfer or dissemination of technology or follow-up action on the part of government and institutions. As we are aware that only specific chemicals should be used, chemicals of unknown origin create a number of potential risks not only to fish but ultimately to the highest consumer, i.e., humans, and impart pollution to both aquatic and terrestrial environments (FAO, 2003). These include:

- Risks to the environment, such as the potential effects of chemicals on water and sediment quality, natural aquatic communities, and effects on microorganisms;
- Risks to human health, such as the dangers to aquaculture workers posed by handling of feed additives, chemotherapeutics, hormones, disinfectants, pesticides, and vaccines, and the risk of developing strains of pathogens that are resistant to antibiotics used in human medicine;
- Risks to production systems for other domesticated species, such as through the development of drug-resistant bacteria that may cause disease in livestock.

As reported, small-scale and marginal hatchery owners and primary actors in the supply chain, are not only illiterate and ignorant, but are deprived of any scientific advice, particularly when they are in difficulty due to outbreaks of uncontrollable disease. This result in a considerable knowledge gap and their being deprived of any assistance and way out; the fish breeders and farmers are very much averse to any government and scientific personnel. This knowledge gap and want of any scientific supervision play a major role in the recent development of a number of negative genetic consequences as envisaged in Chapter 2, Reproductive Cycle, Maturation, and Spawning. Even though poor seed quality is often attributed to inbreeding pressure of the brood stock used, over short timescales, it would actually be very unlikely that inbreeding is a major contributor (Mair, 2002). Present day hatcheries are passing through profit-making approaches along with conversion of native to exotic fish hatchery and many other misleading approaches. The consequences have already been elucidated in Chapter 2, Reproductive Cycle, Maturation, and Spawning. This drift of captive breeding program from its goal of quality seed production has already been established by various studies. Quarantine procedures and practical farm-level biosecurity measures are also lacking.

Research agendas should be developed jointly with farmers so that their needs are addressed appropriately. A number of researchable topics are listed below:

- Tools for decision-making concerning brood stock and seed quality, breeding and culture environment and risk-reduction measures against diseases;
- Indigenous brood stock management, breeding, genetics, and fry nursing practices;
- Use of wild-caught fish as feed for predatory fish;
- Suitable farm-made feeds;
- Integrated Pest Management (IPM) and aquaculture in rice fields.

8.8 Future Prospects and Recommendations

8.8.1 Hatchery and Nursery Management

1. Hatchery and nursery operators of different scales should be made aware of the options and opportunities available to them for controlling diseases and maintaining the quality of brood stock and fish seed. In order to provide practical and effective technical guidance for hatchery and nursery management, it is first necessary to review the basic requirements for an effective hatchery and nursery production.

2. Develop good hatchery and nursery management practices and document them with adequate scientific evidence, and field data are appropriate and timely. Currently, harmonized technical standards/guidelines for hatchery production and fry nursing are lacking. It is important that such technical standards be developed, standardized, validated, and agreed upon by hatchery operators, both nationally and internationally, and by large-scale and small-scale producers.

3. Accreditation of hatcheries at different production scales will provide an opportunity for the recognition of hatcheries and thereby influence the production of high-quality seed.

4. Develop strong management tools for hatchery and nursery operators to facilitate their own decision-making on brood fish and fish seed quality, breeding, and culture environment.

5. Develop breeding technologies of common culture species dependent on wild seed to ease pressure on wild fishery and transboundary movement of large amounts of fish seed.

8.8.2 Capacity Building

1. Farmers have a wealth of local knowledge on fish breeding and nursery rearing. This local knowledge should be studied, assessed, documented, and disseminated.
2. The capacity-building themes for farmers wishing to enter into fish breeding and fry nursing should include indigenous knowledge-based spawning and fry-rearing techniques. Successful minihatchery operators and nursery farmers who have indigenous knowledge should be used as resources for such capacity-building programs.
3. Adopt an incentive scheme for hatchery and fry nursery operators and fry/fingerling traders who have developed good management practices, e.g., using their services for extension. Encourage the spread of such services on a voluntary or paid basis. Make use of them in government training and development programs to utilize their capacities as trainers and facilitators on an incentive/payment basis. This can reduce the cost of extension.
4. Prepare a directory of hatchery/nursery operators practicing good management practices to make their services available at village/commune level and also for development programs. Seed producers and fry traders can be effectively used as trainers since they usually have a good appreciation of what is possible under local conditions.
5. Extension should not only be limited to strengthening knowledge and skills and changing attitudes and behavior to become primary producers but also developing entrepreneurship skills to become successful traders.

8.8.3 Research and Development

1. Research institutes should build an "institute–industry research partnership" with hatchery and nursery operators to improve quality brood stock and to produce quality fingerlings.
2. Undertake research studies to determine the cause(s) of poor-quality seed.
3. Investigate technological innovations to use rice fields for fingerling production.
4. Research support should be channeled to researchable subjects based on farmer needs.
5. Research on the use of wild-caught fish as feed as a basis for support to policy decisions for efficient resource use.

8.8.4 Policy Directions

1. The principal deficiency in the legal framework with respect to aquaculture is that there is no legal definition of aquaculture provided in the sector governing principal acts in almost all

the countries in the region. Aquaculture tends to be considered as part of national fisheries legislation. When a legal definition for aquaculture is prepared, the collateral issues relating to the aquaculture facility or the aquaculture product will be taken into account and covered by the appropriate legislation. Therefore, a legal definition will facilitate the regulation of the fish seed industry, which is not in place in many countries.

2. It appears that there are no reliable statistics and available statistics certainly underestimate the contribution made by large numbers of small-scale household fish seed producers. Reliable statistics can influence policy directions to build effective support services for small-scale aquaculture producers.

3. Translate technical documents and reports produced by development projects into local languages and circulate to concerned institutions and interested parties such as planners and policymakers to help in planning follow-up development actions.

4. Codes of conducts have been adopted by ASEAN on shrimp aquaculture, e.g., the Manual of ASEAN Good Shrimp Farm Management Practices was adopted at the 20th Meeting of ASEAN Ministers of Agriculture and Forestry(AMAF) held in Hanoi, Vietnam, in 1998. ASEAN has also published two other guidelines on fisheries, namely the Manual on Practical Guidelines for the Development of High Health *Penaeus monodon* Brood Stock and the Harmonization of Hatchery Production of *P. monodon* in ASEAN Countries. Considering the socioeconomic importance of freshwater fish seed supply in aquaculture, similar initiatives may be adopted in the freshwater fish seed production.

5. Shift in government role as a fish seed supplier competing with the private sector to genetic conservation would be more beneficial in the long-term viability of the fish seed industry. The government hatcheries with some back-up by large-scale private hatcheries should focus on maintain genetic stocks and brood stock of species to support overcoming the constraints related to genetic quality, such as inbreeding problems and difficulties in breeding some species, faced by small-scale hatcheries due to lack of pond space and brood stock management capacity.

6. The bulk of existing production is from omnivorous and herbivorous species, both indigenous and exotic species. There is a preference in the region for indigenous species, and this should be supported by brood stock management strategies that preserve genetic diversity.

7. The conversion of wetlands and rice fields for fish nurseries for better returns at the expense of livelihood loss due to wetland degradation and wild fisheries and loss of staple food should be discouraged. Staple foods, such as rice, once sear can not be valued.

8. Encourage and involve decentralized seed production and networking for seed supply to reach remote areas through support from local government institutions. Farmers to have access to high-quality fish seed available at the appropriate time for stocking will ensure the smooth flow of products and value along the entire aquaculture value chain. It is also important that local government line agencies place greater emphasis on providing services and focus on the poor isolated farmers or those who do not have access to private hatcheries to ensure that they are not marginalized and have access to quality brood fish.

8.8.5 Regional Issues to Be Addressed

Introductions of fish seed, including genetically improved strains and exotic species, play an important role in the supply of fish seed for aquaculture. Any movement of fish seed among diverse natural and ecological boundaries/watersheds may involve risk associated with disease spread, biodiversity, and management of genetic resources. Issues related to increased cooperation in research, development, and awareness with respect to transboundary issues involving unauthorized introduction of fish and fish seed, prevention of disease spread, and management of genetic resources, should be taken care of.

References

AIT, 2000. Fish Seed Quality in Asia, Fish Seed Quality in Northern Vietnam, State of the System Report. University of Stirling., 23 pp.

Alam, M.A., Akanda, M.S.H., Khan, M.R.R., Alam, M.S., 2002. Comparison of genetic variability, between a hatchery and a river population of Rohu (Labeo rohita) by allozyme electrophoresis. Pakisthan, Journal of Biological science 4 (7), 959–961.

Brian, H., Ross, C., Greer, D., Carolsfled, J. (Eds.), 1998. Action before extinction, An internatiomal conferenceon Conservation of fish Biodiversity. B.C.WOorld Fisheries treat, Ictoria.

Dan, N.C., Luan, T.D., Quy, N.V., 2001. Selective Breeding of Tilapia *Oreochromis niloticus* for Improvement of Growth Performance and Cold Tolerance. Final Technical Report of Ministry of Fisheries funded Research Project. Hanoi, Vietnam, Ministry of Fisheries.

Dwivedi, S.N., Zaidi, G.S., 1983. Development of carp hatcheries in India. Fishing Chimes 3, 1–19.

Edwards, P., Hiep, D.D., Anh, P.M., Mair, G., 2000. Traditional culture of indigenous common carp in rice fields in northern Vietnam: does it have a future role in poverty reduction? World Aquacul. 31 (4), 34–40.

Eknath, A.E., Doyle, R.W., 1985. Indirect selection for growth and life history traits in Indian carp aquaculture. I. Effects of brood fish management. Aquaculture 49, 73–84.

FAO, 2003. Health Management and Biosecurity Maintenance in White Shrimp (*Penaeus vannamei*) Hatcheries in Latin America. FAO Fisheries Technical Paper No.450. FAO, Rome.

FAO, 2005a. Aquaculture Production: Quantities 1950–2004. FISHSTAT Plus universal software for fishery statistical time series. Available at www.fao.org.

FAO, 2005b. FishStat Plus: Fisheries & Aquaculture On-Line Production Statistics. FAO, Rome, Available at http://www.fao.org/fi/statist/FISOFT/FISHPLUS.asp).

Fisheries Law of the People's Republic of China (1986, as amended in 2000).

Hao, N.V., Sang, N.V., Khanh, P.V., 2004. Selective Breeding of Mekong Striped Catfish *Pangasius hypophthalmus*. Final Technical Report of Ministry of Fisheries funded Research Project. Hanoi, Vietnam, Ministry of Fisheries.

Hasan, M.R., Ahmed, G.U., 2002. Issues in carp hatcheries and nurseries in Bangladesh,with special reference to health management. In: Arthur, J.R., Phillips, M.J., Subasinghe, R.P., Reantaso, M.B., MacRae, I.H. (Eds.). Primary Aquatic Animal Health care in rural, Small-Scale, Aquaculture Development. FAO Fish. Tech. Pap. No. 406, pp. 147–164.

Hussain, M.G., Mazid, M.A., 1997. Problems of inbreeding and cross breeding in hatchery and their remedial mitigating measure. In: Hasan, M.R., Rahman, M.M., Sattar, M.A. (Eds.). Quality Assurance in Induced Breeding. Jessore: Department of Fisheries, pp. 7–11.

Ing Kimleang, 2004. The impact of imported fisheries production and fisheries products on Cambodia's fisheries and economy. Royal University of Agriculture, Phnom Penh, Cambodia. (M.Sc. thesis).

Ingthamjitr, S., 1997. Hybrid Catfish Clarias Catfish Seed Production and Marketing in Central Thailand and Experimental Testing of Seed Quality. Asian Institute., Bangkok.

Little, D.C., 1998. Seed quality becomes an issue for fish farmers in Asia. AARM Newsletter 3 (3), 10–11.

Little, D.C., Satapornvanit, A., Edwards, P., 2002. Freshwater fish seed quality in Asia. In: Edwards, P., Little, D.C., Demaine, H. (Eds.), Rural Aquaculture. CABI Publishing, Oxon, pp. 185–193.

Mair, G.C., 2002. Supply of good quality fish seed for sustainable aquaculture. Aquaculture Asia 7 (2), 25–27.

Matsha and Chingri Hatchery Ayen, 2005. Ministry of Fisheries and Livestock has drafted the Law for Fish and Shrimp Hatchery 2005, known as "Matsha and Chingri Hatchery Ayen, 2005".

Mondal, B., Mukhopadhyay, P.K., Rath, S.C., 2005. Bundh breeding of carps: a simple innovative technique from District Bankura, West Bengal (India). Aquacul. Asia 10 (2), 9–10.

Nam, S., Leap, H., 2007. An evaluation of freshwater fish seed resources in Cambodia, pp. 145–170. In: Bondad-Reantaso, M.G. (Ed.), Assessment of freshwater fish seed resources for sustainable aquaculture. FAO Fisheries Technical Paper. No. 501. FAO, Rome. 2007. 628p.

NEFP, 1993–1994. Project monitoring and evaluation report. In: Morrice, C.P. (comp). *Northwest Fisheries* Extension Project, Parbatipur, Annual Report. Bangladesh, Department of Fisheries.

Phillips, M.J., 2002. Freshwater Aquaculture in the Lower Mekong Basin. MRC Technical Paper No. 7. Mekong River Commission, Phnom Penh.

Ramanathan, S., 1969. A preliminary report on *Chanos* fry. Surveys carried out in the brackish water areas of Mannar, Puttalam and Negombo. Bull Fish Res Station Ceylon 2, 79–85.

Simonsen, V., Hansen, M.M., Sarder, M.R.I., Alam, M.S., 2004. High level hybridization in three species of Indian major carps. NAGA 27 (1), 65–69.

Simonsen, V., Hansen, M.M., Sarder, M.R.I., Alam, M.S., 2005. Widespread hybridization among species of Indian major carps in hatcheries but not in the wild. J Fish Biol 67, 794–808.

So, N., Eng, T., Souen, N., Hortle, K.G., 2005. Use of Freshwater Low Value Fish for aquaculture Development in the Cambodia's Mekong Basin. Consultancy report for Mekong River Commission – Assessment of Mekong Capture Fisheries Project. Department of Fisheries, Phnom Penh.

Thayaparan, K., Chakrabarty, R.D., 1984. In: Juario, J.V., Ferraris, R.P., Benitez, L.V. (Eds.), *Advances in Milkfish Biology and Culture: Proceedings of the Second International Milkfish Aquaculture Conference, 4–8 October 1983, Iloilo City, Philippines*, Published by Island Pub. House in association with the Aquaculture Dept., Southeast Asian Fisheries Development Center and the International Development Research Centre.

Tien, T.M., 1993. Review of fish breeding research and practices in Vietnam Proc. of Workshop on selective breeding of fishes in Asia and the United States. Oceanic Inst., Honolulu, HI, May 3–7. 267 pp.

Tien, T.M., Dan, N.C. & Tuan, P.A. 2001.Review of fish genetics & Breeding research in Vietnam.In M.V. Gupta& B.O.Acosta(eds.) . Fish genetics research in member countries ans Institutional network on Genetics in Aquaculture, ICLARM, Conference Proceedings . 64, 179 pp.

Tuan, P.A., Hung, L.Q., Diep, H.T., Tan, N.T., 2005. Investigation Into Genetic Improvement of Grass Carp and Mrigal. Final Technical Report of Ministry of Fisheries funded Research Project. Hanoi, Vietnam, Ministry of Fisheries.

9

REVIEW OF FRESHWATER PRAWN HATCHERY OPERATION IN SOME ASIAN COUNTRIES

9.1 Introduction

This chapter looks at the production and economic performance of freshwater prawn (*Macrobrachium rosenbergii*) hatcheries in Bangladesh. A rigorous investigation was carried out on 15 randomly selected hatcheries in the southwest region of Bangladesh. More than 95% hatcheries were found to use wild brood stock at average individual weight of around 70–150 g. After hatching, the larvae are fed *Artemia* initially and from the 11th day onward custard is provided additionally. Larvae are reared at 12‰ salinity until the post larvae (PL) stage appeared within 33–45 days. The hatcheries conduct two productions cycle per year with each one running for 50 days. The average PL production was 1.72 ± 10.84 millions/cycle. The larval survival rate was very poor 20–35%; viral and bacterial diseases were identified as major cause of larval mortality. Additionally, other hiccups were natural disasters, unexpected weather conditions, lack of credit facility, inadequate power supply, and uncontrolled market conditions. The economic analysis of the present cycle represented a benefit–cost ratio of 1.36 while the net return was estimated at US$7595.44. However, production is seasonal and often altered by sudden climatic disaster, disease outbreak, and a market containing illegally received foreign PL. Therefore, the net return achieved from the hatcheries is not satisfactory for the sustainability of a profitable business.

Induced Fish Breeding. DOI: http://dx.doi.org/10.1016/B978-0-12-801774-6.00009-2

The farming of the giant freshwater prawn (*Macrobrachium rosenbergii* de Man 1879) has been one of the most emerging economic activities in the southwest regions of Bangladesh and over the last two decades its development has been significantly tangible due to its huge export potential (Hasanuzzaman et al., 2011). The prawn and shrimp sector as a whole is the second largest export product after readymade garments from India (DoF, 2013). The southwest part of Bangladesh, where around three-quarters of Bangladesh's prawn farms have been established, appears to be the most important and promising area of prawn culture for the availability of wild PL, with favorable resources and climatic conditions (low lying land, warm weather, fertile soil, and cheap and abundant labor).

Although the resources have enabled a huge expansion of prawn culture, the lack of a stable seed supply has been identified as a tremendous obstacle for further development of this sector (Phuong et al., 2006). Farmers who traditionally depended on wild seed are now facing problem from dwindling natural resources and poor performance (in terms of survival and metamorphosis rates) of larvae from wild captured parent stock which remains a bottleneck (Wilder et al., 1999; Amrit and Yen, 2003). The seed problem has become intense since the late 1980s, due to extensive prawn farming (Ahmed, 2000). The culture is expanding rapidly at 10% per year, which, when combined with declining wild catches and increasing demand for wild PL, has resulted in higher prices for PL. Moreover, indiscriminate fishing of wild PL with high levels of by-catch (i.e., nontarget species caught incidentally) destroys the biodiversity of the coastal ecosystem.

The outrageous impacts on ecosystem have provoked the imposition of restrictions on wild PL collection (Ahmed, 2003). In September 2000, Department of Fisheries in Bangladesh Government imposed a ban on wild PL collection. The rationale for the ban was to protect biodiversity from the harmful effects of intensive PL fishing in the coastal zone (DoF, 2002). In these situations many private and Govt. hatcheries were established to produce quality prawn seed and satisfy the demand. Since then a total 70 prawn hatcheries have been established, of which 17 are Government, and 1.25 billion PL are presently produced annually from these hatcheries (DoF, 2013). Many farmers are currently depending on hatchery produced PL for a cheap rate and available supply (Asaduzzaman et al., 2007).

Although the hatchery PL production system is well documented in the literature, the quality of fry remains a critical concern for prawn farmers and the production performance varies greatly among hatcheries, or even between two distinct cycles in a particular hatchery. The PL production is affected by several factors such as quality of brood stock, climatic conditions, and other incentives (management practices, credit facility, and power supply). The economics of the prawn hatcheries require rigorous study to find out their feasibility as a profitable business industry. The aim of this study was therefore to comprehend the production and economic performance of prawn hatcheries in Bangladesh.

Data related to work principles were noted during the peak production cycle from January to April. The necessary data were gathered from both primary and secondary sources. Primary data were collected through direct observation and using a structured questionnaire on subjects of prawn PL production system, major problems behind development, production, and economic performance of prawn hatcheries. The investigation was done by practical survey on production stages and by interviewing the hatchery workers, technicians, and owners from 15 randomly selected prawn hatcheries (5 hatcheries from each district). To support and evaluate the primary data, secondary data were also collected from different sources such as fisheries research institute, state fisheries department, published books and journals. All data were aggregated, analyzed, and processed using Microsoft Excel database software to understand the present production performance and profitability of prawn hatcheries.

9.2 Results

9.2.1 PL Production Practice

9.2.1.1 Sources of Brood Prawn and Water

More than 95% of brood prawns are collected from nature, mostly from nearby rivers, and the rest from gher (a small area of water, especially used for producing prawn and shrimp). These brood prawns possess an average weight class of 70–180 g. Freshwater was brought from different sources, such as underground water, river canal, rainwater, and ponds, while brine water was collected from the sea with salinity range of 120–200‰.

9.2.1.2 Preparation of Spawning and Larval Rearing Tanks

Most of the hatcheries prepared the spawning and larval rearing tanks as shown in Fig. 9.1.

9.2.1.3 Water Management and Treatment

In most hatcheries in the study areas the water was treated with 5–10 ppm hypochlorite with vigorous aeration. Treated water was neutralized by adding sodium thiosulfate with strong aeration until all chlorine residues were evaporated. Treated water was used 3–5 days after neutralization.

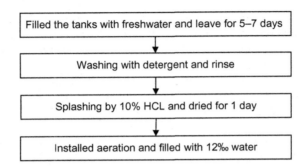

Figure 9.1 Spawning and rearing tank preparation.

9.2.1.4 Brood Stock Management, Selection, and Spawning

Mother prawns were fed with fish meal, rice, shrimp (*Metapenaeus monoceros*), and mollusc flesh twice per day. Matured broods were selected with olive green ovary and opaque white bulging thelyca. Once selected they are disinfected with formalin (0.2 mL/L) and kept in spawning tanks at 25–28°C temperature and 12‰ salinity with continuous aeration. No feed were given to broods during spawning time. After spawning the spent prawns were removed and eggs were allowed to settle at the bottom by turning off the aeration. Eggs were collected and rinsed several times with clean seawater and transferred to the hatching tank. Larvae came out within a few hours under constant aeration.

9.2.1.5 Stocking of Nauplii

The average stocking density of larvae in the larvae rearing tank (LRT) was 106.92 ± 18.43 per liter. The larvae are collected by siphoning them out into a container and transferred to LRT. To determine the larval density per liter (D_i) in the hatching tank the nauplii are counted three times in the subsamples of 100 mL.

The following formula was used to compute the amount of water in each LRT to gain the desired larval density.

$$V = \frac{D_d \times V_L}{D_i}$$

V, volume of water in the container to be placed in the tank
D_d, desired stocking density
V_L, volume of larval tank
D_i, density of nauplii in the container.

Temperature and salinity were adjusted according to the spawning tank.

Table 9.1 Feed Applied for the Larvae in Hatchery

Age (days)	Feed Type	Frequency/ Day	Quantity Increased/Day (g)
1	No feed		
2–10	*Artemia* (instar-1)	Twice	400
11–30	*Artemia* + custard	4 times	600
31–50	*Artemia* + custard	4 times	500

9.2.1.6 Feeding

The surveyed hatcheries provided *Artemia* nauplii from the 2nd to 10th day after hatching (DAH) of prawn larvae and *Artemia* + Custard (prepared from Egg + Corn flower + Golda meat + Milk powder + Vitamin + Tetracycline) was served from the 11th to 50th day, while *Artemia* nauplii were fed only once at night (19.30 h). The amount of feed was increased each day (Table 9.1).

9.2.1.7 Production Performance

The hatchery operators reared larvae up to PL10 to PL15. The larvae reached the PL stage in 35–45 days. The hatcheries of the survey area practice seasonal production from February to July, principally associated with the demand of fry for prawn culture. Most of the hatcheries conduct two cycles per year where the first cycle is done from March to May and the second cycle from June to August. The average PL production was 1.72 ± 10.84 millions/cycle. Among the hatcheries the highest PL production rate was found at BRAC Golda hatchery in Khulna district—about 4 millions/cycle. The hatcheries achieved survival rates of 20–35% from hatching to PL15. The operators sold PL to the farmers and the traders directly.

9.2.1.8 Economics

The economic analysis of the freshwater prawn hatchery production showed that the total annual cost on average was US$20,950.34 $\pm$ 2856.63/cycle of which 40.35% was made up by fixed costs and the rest was variable costs (Table 9.2). The net return from the present cycle was estimated at US$7595.44 (Table 9.3) which is 36.25% of the profit after all operational expenses.

Table 9.2 Total Cost and Income (US$/Production) of Prawn Hatchery

Item		Cost (US$ ± SD)	Useful Life (years)	Depreciation
Capital cost	Land and infrastructure	142,120.07 ± 102,665.75	20	7106.00
	Generator	306.75 ± 55.85	10	30.68
	Sand filter	155.91 ± 45.31	15	10.39
	Total capital cost	142,582.73 ± 81,919.54		7147.07 ± 4090.81
Annual Operating Cost (US$/Production)				
				% TC (Total Cost)
Variable cost	Brine	1581.61 ± 906.34		7.55
	Brood	124.20 ± 17.58		0.60
	Artemia	5391.65 ± 1796.03		25.74
	Artificial diets	2645.40 ± 1796.03		12.63
	Electricity	907.13 ± 297.24		4.33
	Labor and personnel	1448.40 ± 1067.08		6.91
	Water treatment	119.98 ± 37.69		0.57
	Fuel	276.74 ± 185.64		1.32
	Total variable cost (VC)	12,495.11 ± 1776.06		59.65
Fixed cost	Depreciation cost	7147.07 ± 4090.81		34.11
	Repair and maintenance	1022.51 ± 996.08		4.88
	Staff salary	285.65 ± 208.98		1.36
	Total fixed cost (FC)	8455.23 ± 3766.79		40.35
	Total cost (TC) = FC + VC	20,950.34 ± 2856.63		100
	Total income (TI)	28,545.78 ± 22,106.57		

SD, standard deviation.

9.3 Discussion

Most of the hatcheries are in multiple ownerships. As observed in most Asian countries, the likely main source of brood prawns are native habitats, i.e., the major rivers of the relevant countries. After collection, berried females are transported by individual plastic tube, especially during long transport. As indicated by some authors (Ahmed, 2008) 33% of hatcheries use natural brood stock, 13% use

Table 9.3 Economic Return and Benefit–Cost Ratio From Prawn Hatchery

Economic Indicator	Total (US$/Production)
FC	8455.23
VC	12,495.12
TC	20,950.34
TI	28,545.78
Net return (NR = TI − TC)	7595.44
Benefit–cost ratio (BCR) = TI/TC	1.36

farm-raised brood stock, and the remainder (54%) come from both sources although natural sources dominate the brood stock supply chain. This indicates that the prawn and shrimp hatcheries in the Asian countries are still dependent on wild collection although considerable work and money has already invested for standardization. Another hindrance is that the traditional fish breeders and hatchery owners have a firm belief that wild stock is far better in quality (high survival rate) with higher PL production success than that of hatchery raised stock. In Vietnam about half of the hatcheries rely on wild stock (Marcy, 2005). The average weight of mother broods, collected from nature, ranged from 70 to 180 g among hatcheries but Mohanta (2000) reported that broods ranging from 80 to 120 g are suitable for the production of healthy larvae. Again Rao et al. (1994) elucidated that the larvae produced out of larger berried female exhibit better growth and metamorphosis, due to shortening of the larval culture period. In prawn hatcheries berried females are commonly transferred from freshwater to brackish water to improve hatching rates. The feeding schedule of gravid females varies from place to place but the normal practice is to shift to pelleted feed when formulated feed is lacking. Added to this, a variety of feeds, of plant and animal origin, are also in use (Nandlal and Pickering, 2005). Gonadal maturity is determined by observing the olive green along the margin of ovary, an indication of final maturity, and this assessment of maturation acts as a useful tool in the selection of brood fish. The maturity stages are generally classified based on differential color and size of ovary in relation to the carapace cavity (Rao and Tripathy, 1993). The stocking density of larvae ranged from 90 to 100 per liter

(Nandlal and Pickering, 2005). Larvae are stocked in PL rearing tank with optimum salinity 12 ± 1‰. On day 1 after hatching the larvae sustain on egg yolk, and exogenous food was provided DAH 1 only. The popular food used by major hatcheries are *Artemia* seed of Red jungle USA (OSI) brand. Preparation of live feed involves decapsulation of seed. After 10 days, when the larvae have passed through rapid and successive developmental stages with rapid changes in morphology, physiology, and related behavioral changes, there is a rapid change in nutritional requirements. To compensate the rapid shift from one developmental stage to another the developing larvae are served three times additional feed (custard) and once *Artemia*. Realizing the rapid pace in developmental stages, feed quantity increased each day and the rate increment declined from 600 to 500 g after 30 DAH.

Exchange of culture water is undertaken daily at the rate of 30% during the second protozoan sub stage (ZII) to 50% at the mysis stage. Residual chlorine is determined and for every 1 ppm of hypochlorite an equivalent amount of sodium thiosulfate is added in the chlorinated water with vigorous aeration to facilitate neutralization which continues until the residual chlorine becomes zero. PL appear within 33–45 days of rearing (vary among hatcheries) when 12‰ salinity is maintained in some regions, whereas according to Soundarapandian et al. (2009) it takes 27–33 and 29–66 days at 10–15‰ salinity level, respectively, in some of the Asian countries. In Fiji, hatcheries' larvae took 22–25 DAH to attain PL stage under captive condition (Nandlal and Pickering, 2005). In most of the hatcheries in the Asian region the larvae takes a long duration larval stage to reach PL but further work is needed to identify the agents. The mortality rate ranged 65–80%, while in Vietnam it is reduced at 10% (Marcy, 2005). The huge mortality is due to viral and bacterial invasion. The PL production varies from 0.15 to 4 million/cycle among the hatcheries. It is worth mentioning that all the hatcheries attain only one-third of the actual production potential. Low larval survival rate, deficiency in understanding the proper management conditions, along with insufficient investment are identified as the principal criteria behind the varied production performance.

The net profits vary from country to country—while India and Bangladesh registered a profit of US$7595.43, documented profit margin is higher compared to that of Vietnam (US$7000) (Marcy, 2005). Whereas a profit of 36.25% is recorded in India and Bangladesh, it is only 30% for a small-scale hatchery with a target production of 1 million PL/cycle in the Philippines. Although hatchery production is a profitable proposition, oftentimes a setback appears suddenly incurring a total loss. These reasons include natural calamities, unstable public demand, prohibition on prawn export, quality of broods, insufficient funding, chemical impurities, disease outbreak, as well as a scientific knowledge gap and misleading approaches on the part of agents, dealers, and company representative. Uncontrolled weather conditions along with sudden shifts in temperature and associated water quality parameters are the key factors interrupting the cycle. Interruption in electricity supply and a lack of certified seed are the other factors producing negative impacts in hatchery production. Hatchery owners admitted that cross country PL transactions through illegal paths and at a lower price, in the face of a lack of indigenous PCR tested and certified seed, is a major threat for the development of the prawn and shrimp culture sector in most of the Asian countries.

9.4 Conclusion

The prawn hatchery sector, instead of having an inherent quality, can be a well-supported livelihood option for the poor masses. The review's report of cross country culture practices is not encouraging as the production is very much unstable in the absence of positive initiatives from the Government. As the production is seasonal with a maximum of—one to two cycles per year, this acts as bounding factor impeding desired production. Good management training for the workers, ensuring hygienic practices, sufficient credit facilities, and a constant power supply, along with a controlled marketing system are necessary for the sustainability of the hatchery industry. During our review we have noticed dreadful negative profit-making approaches involving rapid conversion of paddy fields into prawn culture and this conversion is not static but continues year after by adopting new paddy fields and leaving the previously adopted fields unused. Biodiversity of the entire area is threatened.

References

Ahmed, N., 2000. Bangladesh is need prawn hatcheries: farmers seek solution to wild fry dependency. Fish Farm. Int. 27, 26–27.

Ahmed, N., 2003. Environmental impacts of freshwater prawn farming in Bangladesh. Shell Fish News 15, 25–28.

Ahmed N., 2008. Freshwater prawn hatcheries in Bangladesh: concern of brood stock. Aquaculture Asaia Magazine 8/10/2008. www.enaca.org/modules/news/article.php?article id=1777.

Asaduzzaman, M., Wahab, M.A., Yi, Y., Diana, J.M., Lin, C.K., 2007. Bangladesh prawn farming survey reports industry evaluation. Global Aquacult. Advocate 9, 40–43.

Amrit, N.B., Yen, P.T., 2003. Comparison of larval performance between Thai and Vietnamese giant freshwater prawn, *Macrobrachium rosenbergii* (de Man): a preliminary study. Aquacult. Res. 34, 1453–1458.

DoF (Department of Fisheries), 2002. Shrimp Aquaculture in Bangladesh: A Vision for the Future. Department of Fisheries (DOF), Ministry of Fisheries and Livestock, Dhaka, Bangladesh, 7 p.

DoF (Department of Fisheries), 2013. Saranica, Matsya Pakhya Sankalan, Ministry of Fisheries and Livestock. The Government of Peoples Republic of Bangladesh, Dhaka, p. 144.

Hasanuzzaman, A.F.Md, Rahman, M.A., Islam, S.S., 2011. Practice and economics of freshwater prawn farming in seasonally saline rice field in Bangladesh. Mesopot. J. Mar. Sci. 26 (1), 69–78.

Marcy, N.W., 2005. Freshwater prawn seed production, culture expand in Vietnam. Ph.D. College of Aquaculture and Fisheries, Cantho University, Cantho City, Vietnam, 66 p.

Mohanta, K.N., 2000. Development of Giant Freshwater Prawn Broodstock. Institute of Freshwater Aquaculture, Kausalyaganga, Bhubaneswar, Orissa, India, 751002, 2 p.

Nandlal, S., Pickering, T., 2005. Freshwater prawn *Macrobrachium rosenbergii* farming in Pacific Island countries. Volume 1. Hatchery operation. Secretariat of the Pacific Community, Suva, Fiji Islands, Noumea, New Caledonia. pp. 12–18.

Phuong, N.T., Hai, T.N., Hien, T.T.T., Bui, T., Huong, D.T.T., Son, V.N., et al., 2006. Current status of freshwater prawn culture in Vietnam and the development and transfer of seed production technology. Fish. Sci. 72, 1–12.

Rao, K.J., Tripathy, S.D., 1993. A manual of giant freshwater prawn hatchery. CIFA Manual Ser. 2, 50.

Rao K.J., Rangacharyulu P.V., Bindu R.P., 1994. A two phase larval rearing system for *M. rosenbergii*. Paper presented at the workshop on Freshwater Prawn Farming in India, March 1994, CIFE, Mumbai.

Soundarapandian, P., Prakash, K.S., Dinakaran, G.K., 2009. Simple technology for the hatchery seed production of Giant Palaemonid Prawn *Macrobrachium rosenbergii* (de Man). J. Animal Vet. Adv. 1 (2), 49–53.

Wilder M.N., Yang W.J., Huong D.T.T., Maeda M., 1999. Reproductive mechanisms in the giant freshwater prawn. *Macrobrachium rosenbergii* and cooperative research to improve seed production technology in the Mekong delta region of Vietnam, UJNR Technical Report No. 28.

10

GENETIC VARIATION AND PHYLOGENETIC RELATIONSHIP AMONG THE TWO DIFFERENT STOCKS OF CATLA (*CATLA CATLA*) IN THE INDIAN STATE OF ORISSA BASED ON RAPD PROFILES

The catla carp (*Catla catla* Hamilton: 1822) is an Indian major carp (IMC) and one of the major aquaculture species of India, Bangladesh, Myanmar (Burma), and Pakistan (Jhingran, 1986). In India, *C. catla* is found in almost all the major and minor rivers, lakes, and beels and it is being cultured in all the states of India due to its high growth rate and high market price.

The natural breeding of *C. catla* has become uncertain due to continuous degradation of habitat caused by environmental modification and manmade interventions, leading to decreased catla populations in the state of Orissa.

Induced Fish Breeding. DOI: http://dx.doi.org/10.1016/B978-0-12-801774-6.00010-9

Inbreeding is a common scenario in all the hatcheries of Orissa with mass stocking of genetically inferior fry in hatcheries having the potential to cause feral gene introgression into hatchery raised seeds. Little genetic information is available on *C. catla* species in Puri and Ganjam Districts of Orissa state. Therefore, this study was carried out to evaluate the genetic variation and phylogenetic relationship among the two different stocks of *C. catla* in Orissa based on RAPD profiles (Fig. 10.1).

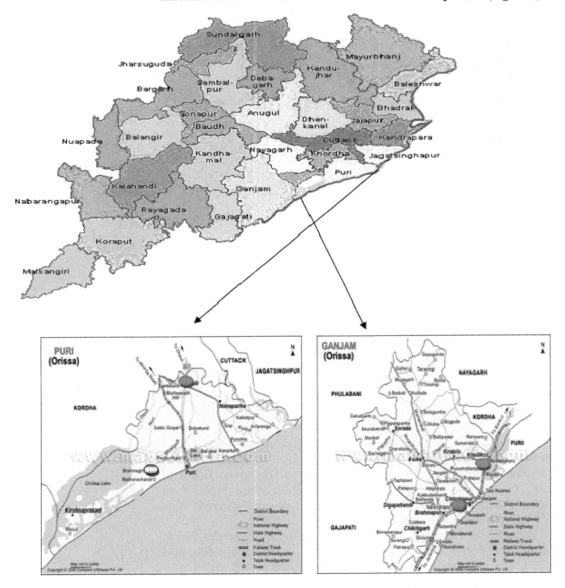

Figure 10.1 Map of Orissa showing sampling sites in the two different districts of Orissa Puri and Ganjam.

In the present investigation, the genetic diversity of two stocks of *C. catla* was studied through RAPD analysis and the phylogenetic cluster was done using UPGMA with NTSYS pc version 2.2 software. The result indicated that the two stocks of catla fall under one phylogenetic cluster. The mean intraspecies genetic similarity (GS) values were 0.7897 ± 0.663 for Puri stock and 0.836 ± 0.599 in the case of Ganjam stock, respectively. The highest GS values within the stock were obtained for Ganjam followed by Puri. These intraspecies GS values estimated for two stock were checked by one-way ANOVA (SPSS version 16) and found to be significantly different at p.

10.1 Introduction

The aquaculture systems in India and its neighbors, such as Bangladesh and Pakistan, mainly constitute IMCs, namely, the catla (*C. catla;* Hamilton), the rohu (*Labeo rohita;* Hamilton), and mrigal (*Cirrhinus mrigal*; Hamilton). These carp contribute approximately 75% to the aquaculture production in India (FAO, 1997). As these carp are economically very important, research on cultivating these species of fish was initiated in India during the early 1950s to study and understand the biology of these economically important species, particularly the major carp, and to develop suitable technologies for various farming systems.

However, as a single species, *C. catla* contributed 1,235,992 mt. to global aquaculture production in 2005 (FAO, 2007). To identify molecular genetic markers, various techniques are used to detect polymorphisms in the population. Polymerase chain reaction-based DNA typing requires no prior knowledge of target DNA sequence, termed "RAPD-PCR" (for random amplified polymorphic DNA) and is found to generate genetic diversity assessment in fishes. The knowledge of genetic background of a species and its population structure is essential for success in breeding, management, and conservation programs in fisheries.

Individuals with a high level of genetic variation have great prospects in aquaculture in terms of higher growth rate, development, stress, and disease resistance (Carvalho, 1993). Therefore, detection of genetic variation at species and population level is of crucial importance for sustainable aquaculture practices. Genetic variation at species level helps to identify the taxonomic units and to determine the species distinctiveness that can provide essential information for conservation, systematic, ecological, and evolutionary studies (Schiverwater et al., 1997).

Variation at the population level can provide an idea about how many different genetic classes are present and the GS among them, how much diversity is present in those classes and their evolutionary relationship with wild relatives. The genetic variability within populations is extremely useful to gather the information on individual identity, breeding pattern, degree of relatedness, and distribution of genetic variation among them.

Such studies are highly informative for the management of a small group in ex situ collection, cultivars identity, breed or clonal identification and for paternity testing. Williams et al. (1991) developed a PCR-based genetic assay named the random amplified polymorphic DNA (RAPD) method, which detects nucleotide sequence polymorphism in genomic DNA regions by using short oligonucleotide primers of arbitrary sequence, typically a length of 10 nucleotides. RAPD assay reports numerous loci that allow the examination of genetic variation and phylogenetic relationship without prior knowledge of DNA sequence.

Since it is PCR-based, RAPD is a relatively easy, fast, and efficient method, which requires only a small amount of DNA for the study. This technique has been useful for the detection of genetic variation in various fish species. It has been used for phylogenetic studies for species and subspecies identification in fish. The RAPD-PCR technique was used by Williams et al. (1990) and Welsh and McClelland (1990) for fish species identification. The present investigation involves the evaluation of the genetic diversity and the molecular phylogeny within and between the two stocks of *C. catla* from Orissa based on RAPD technique which in turn will be useful to provide valuable information for breeding, conservation, systematic, and ecological and evolutionary studies in *C. catla*.

10.2 Materials and Methods

Catla yearlings were collected from four different places in Orissa. A total of 30 samples from each place was collected. The first sample was collected from a private hatchery at Puri near Brahmagiri, the second sample was collected from the Government Hatchery of the State Department of Fisheries, Kausalyaganga, the third sample was collected from Ganjam District at Puroshattampur (Buguda), and the fourth sample was from the same district at Chatrapur. Each sample was collected inside the district at a distance of about 60–70 km from each other. A small piece of fin tissue was cut from the pelvic, caudal, and dorsal parts of the live fish using clean scissors and forceps. The hands of the collectors were cleaned of mucus and scales between handling different fish. The scissors, knife, or forceps were rinsed with water and every time the fins were cut the knife, scissors, or forceps were disinfected with potash/sodium chloride/95% alcohol. The process followed here is wet fin clipping. The fins were cut and placed in a small plastic vial containing high-strength (95%) ethanol to pre-serve the tissues at room temperature.

DNA was isolated from the fish tissue using a standard SDS-phenol/chloroform method with minor modifications (Singh, 2000). About 100–150 mg of tissue was taken in 500 mL cell lysis buffer in a sterile Eppendorf tube. A total of 3–4 mL of Proteinase-K (20 mg/mL) was added to it and mixed thoroughly.

The solution was incubated at 54°C for 5 h (or 37°C overnight) till the completion of cell lysis. An equal amount (i.e., 500 mL) of Tris-saturated phenol and chloroform and isoamyl alcohol (P:C:I) mix-ture (25:24:1) was added and mixed gently by inverting the tube. The above mixture was centrifuged at 10000 rpm for 10 min and the upper aqueous phase was collected in a fresh Eppendorf tube without dis-turbing the interphase.

The phenolization was repeated till the top layer was transparent and clear. The clear aqueous phase was pipetted out and one-tenth volume of 3 M sodium acetate (pH 5.2) solution was added to it, fol-lowed by gentle mixing.

The DNA was precipitated out with two-third volumes of 2-pro-panol (or two volumes of chilled absolute alcohol). The DNA pellet obtained after spinning at 5000 rpm for 5 min at 4°C was washed two or three times with 70% alcohol and finally once with absolute alco-hol. Then it was air-dried.

Finally, DNA pellet was dissolved in 100 mL of TE (10 mM Tris and 1 mM EDTA) buffer and stored at 4°C for further use. RNase treatment was optional and steps 8–10 were repeated. Then the quantification of the obtained DNA was done by Thermo UV-VIS 1 spectrophotom-eter. Optical densities of the DNA samples were measured at 260 and 280 nm to determine the quantity and purity of isolated genomic DNA and then OD_{260}/OD_{280} ratios were also calculated. Then the RAPD analysis was carried out by random primers named OPC-02, OPC-16, OPC-18, and OPC-11 by PCR amplification.

Then the RAPD- PCR products were detected using agarose gel electrophoresis where the staining of gels was done by ethidium bro-mide (EtBr). The molecular weight determination was then done by lab work software and the statistical analysis of the data was carried out. The RAPD fingerprinting generated for all samples were com-pared within and between different populations. Data were scored manually, based on the presence or absence of a band of identical molecular size. The presence of bands was scored and added to get number of bands with a particular primer.

Comparisons were carried out between samples amplified by the same primer in a pair-wise manner. The similarity index between individuals was calculated following the method of Nei and Li (1979).

GS between individuals A and B (S_{AB}) was calculated using the formula:

$$S_{AB} = 2N_{AB}/(N_A + N_B)$$

where,

S_{AB} = rate of band sharing
N_A = number of bands in individual A
N_B = number of bands in individual B
N_{AB} = total number of bands shared by both A and B individual.

Furthermore, the statistical analysis was carried out by using one-way analysis of variance (ANOVA) in SPSS version 16 program to estimate genetic variation. Cluster analysis was then done using UPGMA to create a dendrogram with numerical taxonomy and the multivariate analysis system (NTSYS-pc, version 2.2) program.

10.3 Results

10.3.1 Yield of Genomic DNA

The yield of genomic DNA isolated from screened fish tissue was 10–15 µ g/mg of tissue. Most of the isolated DNA samples were pure and had an OD ratio at 260/280 nm between 1.7 and 1.8. The samples having OD ratio < 1.7 were reextracted with phenol:chloroform:isoa mylalcohol (25:24:1) once and with chloroform:isoamylalcohol (24:1) twice and repeated the DNA precipitation as mentioned in Materials and Methods so as to obtain OD ratio of ≥ 1.7. The DNA sample was treated with RNase to avoid RNA contamination. After extraction and quantification, the quality of isolated DNA was checked in 0.8% agarose gel electrophoresis and finally preserved in TE buffer at − 20°C.

10.3.2 DNA Fingerprinting by RAPD-PCR

10.3.2.1 Screening of Primers

Reaction conditions for PCR were standardized to generate polymorphic pattern. DNA samples of one catla species from two different districts of Orissa were selected randomly from the total of 20 samples for primer screening and selection. A total of four decamer primers, namely OPC-02, OPC-16, OPA-18, and OPA-11 from Operon Technologies Inc. series, were tested for PCR amplification in one species of catla. All primers have shown bands amplification in one species of *C. catla* collected from two different districts of Orissa but out of these four primers, all primers generated very strong amplification with 75–80% reproducible polymorphic DNA bands in the above tested two stocks. Therefore, four primers, namely, OPC-02, OPC-16,

OPA-18, and OPA-11, were finally selected for polymorphic DNA analysis in the present investigation.

10.3.2.2 Detection of RAPD Products

Electrophoresis on agarose gels and EtBr staining was usually found sufficient for detection of RAPD-PCR products. The RAPD banding profiles (size in bp) were analyzed in PC using "Kodak 1D Image Analysis Software" for the determination of molecular size of amplified bands by comparing with known standard molecular weight markers (100 bp DNA ladder) in the gel. However, the molecular size of DNA bands given in the table after software analysis may vary in a range of 5–10 bp for each molecular size of DNA bands.

10.3.2.3 Intraspecies DNA Band Size Analysis After RAPD-PCR

The molecular weight sizes of the DNA bands generated out of RAPD-PCR in 10 individuals of *C. catla* with each primer are as follows. With OPC-02, RAPD analysis exhibited the banding profile between 320–1300 bp in the case of Puri stock and 280–1500 bp in the case of Ganjam stock (Tables 10.1 and 10.2 and Figs. 10.2 and 10.7). With OPC-16, the Puri samples showed DNA bands in a range of 400–1300 bp and also 400–1300 bp in the case of Ganjam samples, respectively (Tables 10.3 and 10.4 and Figs. 10.3 and 10.8). With OPA-18, the Puri samples showed DNA bands in the range of 230–1400 bp and 230–1400 bp in the case of Ganjam samples, respectively (Tables 10.3 and 10.5 and Figs. 10.4 and 10.9). With Primer OPA-11, the Puri samples showed DNA band size in the range of 280–1400 bp and in the case of Ganjam samples it showed 280–1400 bp (Tables 10.4 and 10.6 and Figs. 10.5 and 10.10). The Puri stock showed about 42.8% polymorphic DNA bands, whereas it was 43.75% in the case of Ganjam samples with OPC-02 (Table 10.7). Primer OPC-16 generated about 75% polymorphic DNA bands in the case of the Puri stock, whereas the Ganjam stock showed 50% polymorphism (Table 10.8). The primer OPA-18 produced about 54.5% of polymorphic bands in the case of Puri stock, whereas it was 50% in the case of Ganjam stock (Table 10.9). Primer OPA-11 generated about 23.07% polymorphic DNA bands in the case of Puri samples and 50% for Ganjam samples (Table 10.8).

10.3.2.4 Interspecies DNA Band Size Analysis After RAPD-PCR

Five DNA samples from each stock were made two representative DNA samples by mixing the two DNA samples into one. The two representative DNA samples from each species were selected for comparative DNA band size analysis among the two stocks of *C. catla* in Orissa (Figs. 10.12–10.15). The primer OPC-02 amplified DNA bands in a range of 280–1400 bp (Table 10.10), whereas OPC-16, OPA-18,

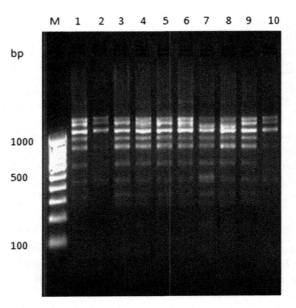

Figure 10.2 RAPD profile of 10 individuals of *Catla catla* collected from Puri using primer OPC-02. Lane M—100 bp DNA molecular weight marker. Lanes 1–10—PCR amplified product of 10 individuals of *Catla catla*.

and OPA-11 amplified 400–1300, 230–1400, and 280–1400 bp between the two DNA samples of two stocks respectively (Tables 10.9–10.12). About 33.33% polymorphic DNA bands were revealed by OPC-02 in two stocks of *Catla*, whereas primer OPC-16, OPA-18, and OPA-11 generated 44.44%, 22.22%, and 25% of polymorphic DNA bands, respectively (Tables 10.13 and 10.14).

10.3.2.5 Primer-Wise RAPD Profile

RAPD Profile With Primer OPC-02

The primer OPC-02 yielded DNA bands in the size range of 280–1400 bp (Tables 10.1 and 10.2) with three polymorphic DNA bands (Table 10.15) in the two stocks of *C. catla*. The level of polymorphic pattern within the species revealed by OPC-02 varies among the two stocks studied. The Ganjam stock revealed the highest polymorphism of 43.75%, whereas the Puri stock showed 42.8% within the species (Table 10.18). On an average 43.27% polymorphic DNA bands were obtained by OPC-02 in the two stocks. This primer amplified two scorable DNA bands of size 1100 and 1200 bp in the two stocks of *C. catla* in Puri and Ganjam districts (Fig. 10.13 and Table 10.10) that may be of diagnostic value.

Table 10.1 Size of Each Amplified DNA Band in the Individuals of *Catla catla* With Primer OPC-02 of Puri District

Lanes Rows	Lane 1 (mol.wt.)	Lane 2 (mol.wt.)	Lane 3 (mol.wt.)	Lane 4 (mol.wt.)	Lane 5 (mol.wt.)	Lane 6 (mol.wt.)	Lane 7 (mol.wt.)	Lane 8 (mol.wt.)	Lane 9 (mol.wt.)	Lane 10 (mol.wt.)	Lane 11 (mol.wt.)
Row 1	–	1300	1300	1300	1300	1300	1300	1300	1300	1300	1300
Row 2	–	1200	1200	1200	1200	1200	1200	1200	1200	1200	1200
Row 3	–	1100	1100	1100	1100	1100	1100	1100	1100	1100	1100
Row 4	1000	1000	–	1000	1000	1000	1000	1000	1000	1000	–
Row 5	900	900	–	900	900	900	900	900	900	900	–
Row 6	800	–	–	–	–	–	–	–	–	–	–
Row 7	700	–	–	–	–	–	–	–	–	–	–
Row 8	600	600	–	600	600	600	600	600	600	600	–
Row 9	500	–	–	–	–	–	–	–	–	–	–
Row 10	–	420	420	420	420	420	420	420	420	420	420
Row 11	400	–	–	–	–	–	–	–	–	–	–
Row 12	–	320	320	320	320	320	320	320	320	320	320
Row 13	300	–	–	–	–	–	–	–	–	–	–
Row 14	200	–	–	–	–	–	–	–	–	–	–
Row 15	100	–	–	–	–	–	–	–	–	–	–

Molecular weight standard: 100 bp ladder. Molecular weight unit: bp.

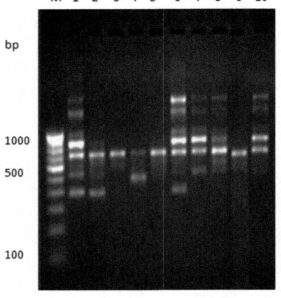

Figure 10.3 RAPD profile of 10 individuals of *Catla catla* collected from Puri using primer OPC-16. Lane M—100 bp DNA molecular weight marker. Lanes 1–10—PCR amplified product of 10 individuals of *Catla catla*.

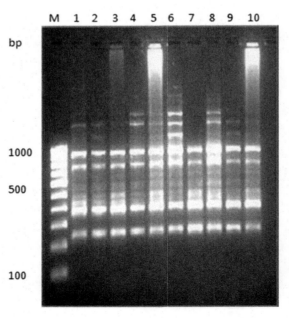

Figure 10.4 RAPD profile of 10 individuals of *Catla catla* collected from Puri using primer OPA-18. Lane M—100 bp DNA molecular weight marker. Lanes 1–10—PCR amplified product of 10 individuals of *Catla catla*.

Table 10.2 Genetic Similarity Values Between 10 Individuals of *Catla catla* of Puri District

Rows/Cols	C. catla1	C. catla2	C. catla3	C. catla4	C. catla5	C. catla6	C. catla7	C. catla8	C. catla9	C. catla10
C. catla	1.000									
C. catla	0.792	1.000								
C. catla	0.772	0.696	1.000							
C. catla	0.780	0.667	0.846	1.000						
C. catla	0.839	0.706	0.873	0.912	1.000					
C. catla	0.875	0.679	0.772	0.881	0.871	1.000				
C. catla	0.807	0.696	0.840	0.692	0.727	0.772	1.000			
C. catla	0.909	0.691	0.814	0.820	0.844	0.848	0.814	1.000		
C. catla	0.750	0.800	0.898	0.824	0.815	0.750	0.816	0.793	1.000	
C. catla	0.800	0.773	0.792	0.680	0.717	0.764	0.833	0.772	0.723	1.000

Values were calculated using software NTSYS-PC.

Table 10.3 Size of Each Amplified DNA Band in the Individuals of *Catla catla* With Primer OPC-16 of Puri District

Lanes / Rows	Lane 1 (mol.wt.)	Lane 2 (mol.wt.)	Lane 3 (mol.wt.)	Lane 4 (mol.wt.)	Lane 5 (mol.wt.)	Lane 6 (mol.wt.)	Lane 7 (mol.wt.)	Lane 8 (mol.wt.)	Lane 9 (mol.wt.)	Lane 10 (mol.wt.)	Lane 11 (mol.wt.)
Row 1	–	1300	–	–	–	–	1300	1300	1300	–	1300
Row 2	–	1200	–	–	–	–	1200	1200	1200	–	1200
Row 3	–	–	–	–	–	1100	–	–	–	–	–
Row 4	1000	1000	–	–	–	–	1000	1000	1000	–	1000
Row 5	900	900	900	–	–	–	900	900	900	–	900
Row 6	800	–	–	–	–	–	–	–	–	–	–
Row 7	–	750	750	750	750	750	750	750	750	750	750
Row 8	700	–	–	–	–	–	–	–	600	–	–
Row 9	600	600	600	–	–	–	–	–	–	–	–
Row 10	–	550	–	–	–	–	–	–	–	–	–
Row 11	500	–	–	–	500	–	–	500	–	–	–
Row 12	400	400	400	–	–	–	400	–	–	–	–
Row 13	300	–	–	–	–	–	–	–	–	–	–
Row 14	200	–	–	–	–	–	–	–	–	–	–
Row 15	100	–	–	–	–	–	–	–	–	–	–

Molecular weight standard: 100 bp ladder. Molecular weight unit: bp.

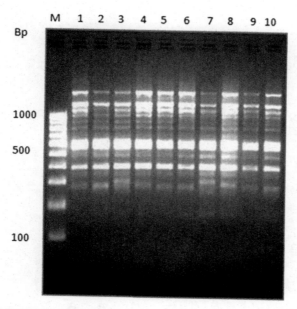

Figure 10.5 RAPD profile of 10 individuals of *Catla catla* collected from Puri using primer OPA-11. Lane M—100 bp DNA molecular weight marker. Lanes 1–10—PCR amplified product of 10 individuals of *Catla catla*.

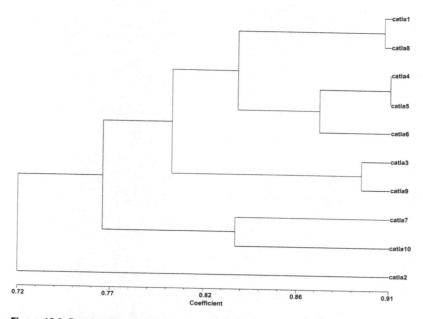

Figure 10.6 Dendrogram of 10 individuals of *Catla catla* collected from Puri District.

Table 10.4 Size of Each Amplified DNA Band in the Individuals of *Catla catla* With Primer OPC-02 of Ganjam District

Lanes Rows	Lane 1 (mol.wt.)	Lane 2 (mol.wt.)	Lane 3 (mol.wt.)	Lane 4 (mol.wt.)	Lane 5 (mol.wt.)	Lane 6 (mol.wt.)	Lane 7 (mol.wt.)	Lane 8 (mol.wt.)	Lane 9 (mol.wt.)	Lane 10 (mol.wt.)	Lane 11 (mol.wt.)
Row 1	–	–	–	–	–	–	–	1500	1500	–	–
Row 2	–	–	–	–	1400	1400	1400	1400	1400	1400	1400
Row 3	–	–	–	–	1350	1350	1350	1350	1350	1350	1350
Row 4	–	1300	1300	1300	1300	1300	1300	1300	1300	1300	1300
Row 5	–	1200	1200	1200	1200	1200	1200	1200	1200	1200	1200
Row 6	–	1100	1100	1100	1100	1100	1100	1100	1100	1100	1100
Row 7	1000	1000	1000	1000	1000	1000	1000	1000	1000	1000	1000
Row 8	900	900	900	900	900	900	900	900	900	900	900
Row 9	800	–	–	–	–	–	–	–	–	–	–
Row 10	700	–	–	–	–	–	–	–	–	700	–
Row 11	600	600	600	600	600	600	600	600	600	600	600
Row 12	500	–	–	–	–	–	–	–	–	–	–
Row 13	–	420	420	420	420	420	420	420	420	420	420
Row 14	400	–	–	–	–	–	–	–	–	–	–
Row 15	–	320	320	320	320	320	320	320	320	320	320
Row 16	300	–	–	–	–	–	–	–	–	–	–
Row 17	–	280	280	280	280	280	280	280	280	280	280
Row 18	200	–	–	–	–	–	–	–	–	–	–
	100	–	–	–	–	–	–	–	–	–	–

Molecular weight standard: 100 bp ladder. Molecular weight unit: bp.

Table 10.5 Size of Each Amplified DNA Band in the Individuals of *Catla catla* With Primer OPC-16 of Ganjam District

Lanes Rows	Lane 1 (mol.wt.)	Lane 2 (mol.wt.)	Lane 3 (mol.wt.)	Lane 4 (mol.wt.)	Lane 5 (mol.wt.)	Lane 6 (mol.wt.)	Lane 7 (mol.wt.)	Lane 8 (mol.wt.)	Lane 9 (mol.wt.)	Lane 10 (mol.wt.)	Lane 11 (mol.wt.)
Row 1	–	–	–	–	–	1300	1300	1300	1300	1300	–
Row 2	–	1200	1200	1200	1200	1200	1200	1200	1200	1200	1200
Row 3	–	1100	1100	1100	1100	1100	1100	1100	1100	1100	1100
Row 4	1000	–	–	–	–	1000	–	–	1000	–	–
Row 5	900	900	900	900	900	900	900	900	900	900	900
Row 6	800	–	–	–	–	–	–	–	–	–	–
Row 7	–	750	750	750	750	750	750	750	750	750	750
Row 8	700	–	–	–	–	–	–	–	–	–	–
Row 9	600	600	600	600	600	600	600	600	600	600	600
Row 10	–	–	–	–	550	550	550	550	550	550	550
Row 11	500	–	–	–	500	–	–	–	500	–	–
Row 12	400	–	400	–	–	–	–	–	–	400	–
Row 13	300	–	–	–	–	–	–	–	–	–	–
Row 14	200	–	–	–	–	–	–	–	–	–	–
Row 15	100	–	–	–	–	–	–	–	–	–	–

Molecular weight standard: 100 bp ladder. Molecular weight unit: bp.

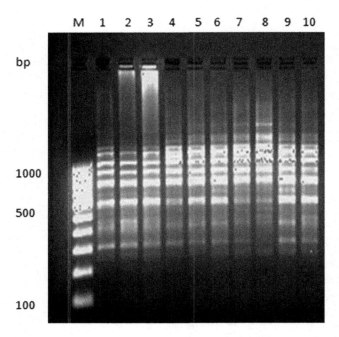

Figure 10.7 RAPD profile of 10 individuals of *Catla catla* collected from Ganjam using primer OPC-02. Lane M—100 bp DNA molecular weight marker. Lanes 1–10—PCR amplified product of 10 individuals of *Catla catla*.

RAPD Profile With Primer OPC-16

The primer OPC-16 yielded DNA bands in the size range of 400–1300 bp (Tables 10.3 and 10.4) with four polymorphic bands (Table 10.15) in the two stocks of *C. catla*. The Puri stock showed the highest polymorphism of 75% with primer OPC-16 (Table 10.14). On an average 62.5% polymorphic DNA bands were obtained by OPC-16 in the two stocks, which was the highest. This primer amplified one scorable DNA band of size 750 bp in Puri stock and Ganjam stock (Figs. 10.3 and 10.8 and Tables 10.3 and 10.4) which can be act as a DNA marker.

RAPD Profile With Primer OPA-18

The primer OPA-18 yielded DNA bands in the size range of 230–1400 bp (Tables 10.6 and 10.16) with two polymorphic bands (Table 10.14) in the two stocks. The polymorphism with primer OPA-18 was 54.5% in the Puri stock, whereas in the case of the Ganjam samples it was 50% (Table 10.14). The average polymorphism shown by this primer among the two stocks is

Table 10.6 Size of Each Amplified DNA Band in the Individuals of *Catla catla* With Primer OPA-18 of Ganjam District

Lanes Rows	Lane 1 (mol.wt.)	Lane 2 (mol.wt.)	Lane 3 (mol.wt.)	Lane 4 (mol.wt.)	Lane 5 (mol.wt.)	Lane 6 (mol.wt.)	Lane 7 (mol.wt.)	Lane 8 (mol.wt.)	Lane 9 (mol.wt.)	Lane 10 (mol.wt.)	Lane 11 (mol.wt.)
Row 1	–	–	–	1400	–	1400	1400	–	–	–	–
Row 2	–	1300	1300	1300	–	1300	1300	–	–	–	–
Row 3	–	–	–	–	–	–	–	–	1100	–	–
Row 4	1000	–	–	–	–	–	–	–	–	–	–
Row 5	–	920	920	920	920	920	920	920	920	920	920
Row 6	900	–	–	–	–	–	–	–	–	–	–
Row 7	800	–	–	–	–	–	–	–	–	–	–
Row 8	–	780	780	780	780	780	780	780	780	780	780
Row 9	700	–	–	–	–	–	–	–	–	–	–
Row 10	600	–	–	–	600	600	600	–	–	–	–
Row 11	500	–	–	–	–	500	500	500	500	500	–
Row 12	400	–	–	–	–	–	–	–	–	–	–
Row 13	–	380	380	380	380	–	–	380	380	380	380
Row 14	300	–	–	–	–	–	–	–	–	–	–
Row 15	–	230	230	230	230	230	230	230	230	230	230
Row 16	200	–	–	–	–	–	–	–	–	–	–
Row 17	100	–	–	–	–	–	–	–	–	–	–

Molecular weight standard: 100 bp ladder. Molecular weight unit: bp.

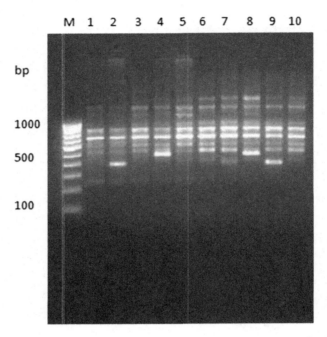

Figure 10.8 RAPD profile of 10 individuals of *Catla catla* collected from Ganjam using primer OPC-16. Lane M—100 bp DNA molecular weight marker. Lanes 1–10—PCR amplified product of 10 individuals of *Catla catla*.

52.25%. This primer amplified two scorable DNA bands of size 230 and 920 bp in the case of Puri stock and 280 and 900 bp in the Ganjam stock (Figs. 10.4 and 10.9 and Tables 10.6 and 10.16) that may be of diagnostic value.

RAPD Profile With Primer OPA-11

The primer OPA-11 yielded DNA bands in the size range of 280–1400 bp (Tables 10.5 and 10.17) with two polymorphic bands (Table 10.14) in the two stocks of catla. In *the* case of Ganjam stock it revealed polymorphism of 50% with primer OPA-11 whereas in the Puri stock it showed 23.03% polymorphism within the species (Table 10.7). The average polymorphism revealed by primer OPA-11 among the two stocks was 36.53%, which was the lowest. This primer amplified four scorable DNA bands of size 400, 550, 620, and 1200 bp in case of Puri stock and was same in the case of the Ganjam stock (Figs. 10.5 and 10.15 and Table 10.9) that may be of diagnostic value.

Table 10.7 Summary of the Number of Total Bands, Common Bands, and Polymorphic Bands of PCR Amplified Products of Two Stocks of *Catla catla* Generated From Four Decamer Primers of Random Sequences

Sl. no.	Species	Primer Code	Total No. of DNA Bands	Common Bands	Polymorphic DNA Bands	Percentage of Polymorphism (%)
1.	Puri	OPA-11	13	10	3	23.07
		OPA-18	11	05	6	54.50
		OPC-02	10	07	3	42.80
		OPC-16	08	02	6	75.00
2.	Ganjam	OPA-11	04	02	2	50.00
		OPA-18	08	04	4	50.00
		OPC-02	16	09	7	43.75
		OPC-16	08	04	4	50.00

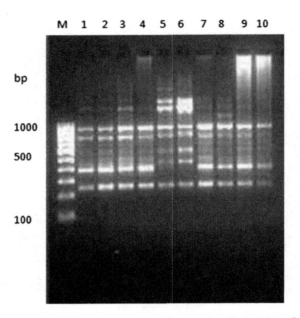

Figure 10.9 RAPD profile of 10 individuals of *Catla catla* collected from Ganjam using primer OPA-18. Lane M—100 bp DNA molecular weight marker. Lanes 1–10— PCR amplified product of 10 individuals of *Catla catla*.

10.3.3 Genetic Polymorphism

The four primers amplified a total of 78 DNA bands ranging in size between 230 and 1500 bp (Tables 10.4 and 10.16) in the two stocks of catla. The total amplified bands ranged between 36–42 and average number was estimated as 39, with four primes in each of the two stocks (Table 10.18). Forty-two of these bands were monomorphic (53.84%) and the remaining 36 bands were polymorphic (Table 10.18). The number of RAPD bands generated per primer ranges between 9.0–10.5, with an average of 9.75 in each stock (Table 10.17). Only reproducible and explicit bands were scored for their presence or absence in the gel (Figs. 10.2–10.15). The percentage of polymorphic bands appeared to be more in the Ganjam stock (47.2%) followed by Puri stock of 42.85 (Table 10.18).

10.3.4 Intraspecies Genetic Variability

The GS matrices for intraspecies variation were calculated with DICE coefficient with NTSYS software by pair-wise comparison of 10 individuals in each stock following the method of Nei and Li (1979). The GS values for each stock ranged from 0.67 ± 0.663 in Puri and

Table 10.8 Similarity Matrix Between the Two Stocks of *Catla catla* Based on Polymorphic DNA Bands

Rows/Cols.	Puri	Ganjam
Puri	1.000	
Ganjam	0.943	1.000

Table 10.9 Size of Each Amplified DNA Band in One sp. of *Catla* With Primer OPA-11 of Two Different Stocks

Lanes Rows	Marker (mol.wt.)	Puri (mol.wt.)	Ganjam (mol.wt.)
Row 1	–	1400	1400
Row 2	–	1200	1200
Row 3	–	1100	1100
Row 4	1000	1000	1000
Row 5	900	–	–
Row 6	800	–	–
Row 7	700	–	–
Row 8	600	600	600
Row 9	–	550	550
Row 10	500	500	500
Row 11	400	400	400
Row 12		320	320
Row 13	300	–	–
Row 14	–	280	280
Row 15	200	–	–
Row 16	100	–	–

Molecular weight standard: 100 bp ladder. Molecular weight unit: bp.

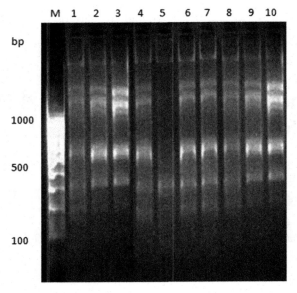

Figure 10.10 RAPD profile of 10 individuals of *Catla catla* collected from Ganjam using primer OPA-11. Lane M—100 bp DNA molecular weight marker. Lanes 1–10—PCR amplified product of 10 individuals of *Catla catla*.

Table 10.10 Size of Each Amplified DNA Band in One sp. of *Catla* With Primer OPC-02 of Two Different Stocks

Lanes Rows	Marker (mol.wt.)	Puri (mol.wt.)	Ganjam (mol.wt.)
Row 1	–	–	1400
Row 2	–	1300	1300
Row 3	–	1200	1200
Row 4	–	1100	1100
Row 5	1000	–	
Row 6	900	900	900
Row 7	800	–	–
Row 8	700	–	–
Row 9	600	600	600
Row 10	500	–	–
Row 11	–	420	420
Row 12	400	–	–
Row 13	300	–	–
Row 14	–	280	280
Row 15	200	–	–
Row 16	100	–	–

Molecular weight standard: 100 bp ladder. Molecular weight unit: bp.

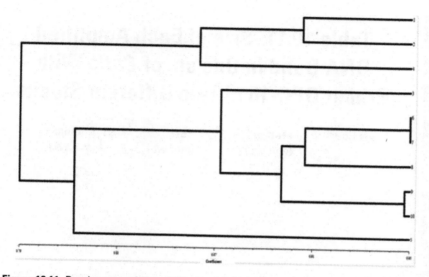

Figure 10.11 Dendrogram of 10 individuals of *Catla catla* collected from Ganjam District.

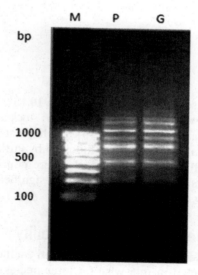

Figure 10.12 Comparative RAPD profile of two different stocks of *Catla catla* using primer OPC-02. Lane M—100 bp DNA molecular weight marker. Lanes P and G— PCR amplified product of two stocks of *Catla catla*.

Table 10.11 Size of Each Amplified DNA Band in One sp. of *Catla* With Primer OPC-16 of Two Different Stocks

Lanes Rows	Marker (mol.wt.)	Puri (mol.wt.)	Ganjam (mol.wt.)
Row 1	–	–	1300
Row 2	–	1200	1200
Row 3	1000		1100
Row 4	900	900	900
Row 5	800	–	–
Row 6	–	750	750
Row 7	700	–	–
Row 8	600	600	600
Row 9	–	550	550
Row 10	500	500	500
Row 11	400	400	400
Row 12	300	–	–
Row 13	200	–	–
Row 14	100	–	–

Molecular weight standard: 100 bp ladder. Molecular weight unit: bp.

0.70 ± 0.599 in case of Ganjam stock (Table 10.15). The mean intraspecies GS values were 0.7897 ± 0.663 for Puri stock and 0.836 ± 0.599 in the case of Ganjam stock (Table 10.15). The highest GS values within the stock were obtained for Ganjam followed by Puri. These intraspecies GS values estimated for two stocks were checked by one-way ANOVA (SPSS version 16) and found to be significantly different at $p < 0.001$ (Table 10.19).

10.3.5 Interspecies Genetic Variability

The two representative DNA samples from each of the two stocks of catla were selected primer-wise for interspecies genetic variability analysis (Figs. 10.12–10.15). The matching coefficients of GS for the same four primers that amplified 78 bands were calculated using

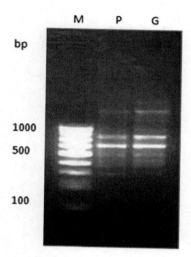

Figure 10.13 Comparative RAPD profile of two different stocks of *Catla catla* using primer OPC-16. Lane M—100 bp DNA molecular weight marker. Lanes P and G—PCR amplified product of two stocks of *Catla catla*.

Table 10.12 Size of Each Amplified DNA Band in One sp. of *Catla* With Primer OPA-18 of Two Different Stocks

Lanes Rows	Marker (mol.wt.)	Puri (mol.wt.)	Ganjam (mol.wt.)
Row 1	–	1400	1400
Row 2	–	1300	1300
Row 3	–	1100	1100
Row 4	1000	–	–
Row 5	900	900	900
Row 6	800	–	–
Row 7	–	750	750
Row 8	700	–	–
Row 9	600	–	–
Row 10	500	500	500
Row 11	400	400	400
Row 12	–	380	380
Row 13	300	–	–
Row 14	–	230	230
Row 1	200	–	–
Row 15	100	–	–

Molecular weight standard: 100 bp ladder. Molecular weight unit: bp.

Table 10.13 Genetic Similarity Values Between 10 Individuals of *Catla catla* of Ganjam District

Rows/Cols	C. catla1	C. catla2	C. catla3	C. catla4	C. catla5	C. catla6	C. catla7	C. catla8	C. catla9	C. catla10
C. catla	1.000									
C. catla	0.905	1.000								
C. catla	0.792	0.840	1.000							
C. catla	0.884	0.844	0.863	1.000						
C. catla	0.727	0.696	0.846	0.766	1.000					
C. catla	0.792	0.800	0.893	0.784	0.885	1.000				
C. catla	0.816	0.842	0.877	0.808	0.830	0.947	1.000			
C. catla	0.800	0.769	0.862	0.868	0.815	0.897	0.915	1.000		
C. catla	0.776	0.784	0.842	0.769	0.755	0.877	0.931	0.881	1.000	
C. catla	0.783	0.792	0.889	0.816	0.760	0.889	0.909	0.893	0.945	1.000

Values were calculated using software NTSYS-PC.

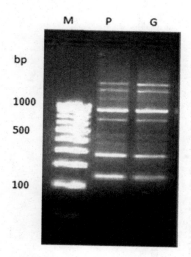

Figure 10.14 Comparative RAPD profile of two different stocks of *Catla catla* using primer OPA-18. Lane M—100 bp DNA molecular weight marker. Lanes P and G— PCR amplified product of two stocks of *Catla catla*.

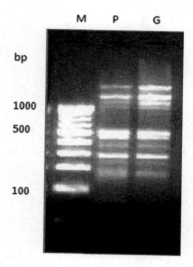

Figure 10.15 Comparative RAPD profile of two different stocks of *Catla catla* using primer OPA-11. Lane M—100 bp DNA molecular weight marker. Lanes P and G— PCR amplified product of two stocks of *Catla catla*.

Table 10.14 Summary of the Number of Total Bands, Common Bands, and Polymorphic Bands and Number of Bands Amplified per Primer of PCR Amplified Products Between Two Stocks of *Catla catla* Generated From Four Decamer Primers of Random Sequences

Sl. No.	Primer Code	Total No. of DNA Bands	Common Bands	Polymorphic DNA Bands	Percentage of Polymorphism (%)
1.	OPC-02	9	6	3	33.33
2.	OPC-16	9	5	4	44.44
3.	OPA-11	9	7	2	22.22
4.	OPA-18	8	6	2	25.00

Table 10.15 Summary of the Mean Genetic Similarity (GS) Values, Standard Deviation, Standard Errors, Minimum and Maximum Values Obtained Within Each of the Species of *Catla* Using SPSS (Version 16) Program

Sl. No.	Species	Mean GS Value	Standard Deviation	Standard Error	Minimum Value	Maximum Value
1.	Puri	0.7897	0.6638	0.00990	0.67	0.91
2.	Ganjam	0.8368	0.5995	0.00894	0.70	0.95

Figure 10.16 Dendrogram of two stocks of of *Catla catla* collected from Ganjam and Puri Districts.

Table 10.16 Size of Each Amplified DNA Band in the Individuals of *Catla catla* With Primer OPA-18 of Puri District

Lanes Rows	Lane 1 (mol.wt.)	Lane 2 (mol.wt.)	Lane 3 (mol.wt.)	Lane 4 (mol.wt.)	Lane 5 (mol.wt.)	Lane 6 (mol.wt.)	Lane 7 (mol.wt.)	Lane 8 (mol.wt.)	Lane 9 (mol.wt.)	Lane 10 (mol.wt.)	Lane 11 (mol.wt.)
Row 1	–	–	–	1400	1400	1400	–	1400	–	–	–
Row 2	–	1300	1300	1300	1300	1300	1300	–	1300	1300	–
Row 3	–	–	–	–	–	–	1100	–	–	–	–
Row 4	1000	–	–	–	–	–	–	–	–	–	–
Row 5	–	920	920	920	920	920	920	920	920	920	920
Row 6	900	–	–	–	–	–	–	–	–	–	–
Row 7	800	800	800	800	800	800	800	800	800	800	800
Row 8	700	–	–	–	–	–	–	–	–	–	–
Row 9	600	–	–	–	–	–	–	–	–	–	–
Row 10	500	–	–	500	–	500	–	500	–	–	500
Row 11	400	–	–	–	–	–	–	–	–	–	–
Row 12	–	380	380	380	380	380	380	380	380	380	380
Row 13	300	–	–	–	–	–	–	–	–	–	–
Row 14	–	230	230	230	230	230	230	230	230	230	230
Row 15	200	–	–	–	–	–	–	–	–	–	–
Row 16	100	–	–	–	–	–	–	–	–	–	–

Molecular weight standard: 100 bp ladder. Molecular weight unit: bp.

Table 10.17 Size of Each Amplified DNA Band in the Individuals of *Catla catla* With Primer OPA-11 of Ganjam District

Lanes / Rows	Lane 1 (mol.wt.)	Lane 2 (mol.wt.)	Lane 3 (mol.wt.)	Lane 4 (mol.wt.)	Lane 5 (mol.wt.)	Lane 6 (mol.wt.)	Lane 7 (mol.wt.)	Lane 8 (mol.wt.)	Lane 9 (mol.wt.)	Lane 10 (mol.wt.)	Lane 11 (mol.wt.)
Row 1	—	1400	1400	1400	1400	1400	1400	1400	1400	1400	1400
Row 2	—	1200	1200	1200	1200	—	1200	1200	1200	1200	1200
Row 3	—	1100	1100	1100	1100	—	1100	1100	1100	1100	1100
Row 4	1000	—	—	1000	—	—	—	—	—	1000	—
Row 5	900	—	—	—	—	—	—	—	—	—	—
Row 6	800	—	—	—	—	—	—	—	—	—	—
Row 7	700	—	—	—	—	—	—	—	—	—	—
Row 8	—	620	620	620	620	—	620	620	620	620	620
Row 9	—	550	550	550	550	—	550	550	550	550	550
Row 10	500	—	—	—	—	—	—	—	—	—	—
Row 11	400	—	—	—	—	—	—	—	—	—	—
Row 12	—	320	320	320	320	320	320	320	320	320	320
Row 13	300	—	—	—	—	—	—	—	—	—	—
Row 14	—	—	—	—	280	280	280	280	280	—	—
Row 15	200	—	—	—	—	—	—	—	—	—	—
Row 16	100	—	—	—	—	—	—	—	—	—	—

Molecular weight standard: 100 bp ladder. Molecular weight unit bp.

Table 10.18 Summary of the Number of Total Bands, Polymorphic Bands, Percentage of Polymorphic Bands and Number of Amplified Bands per Primer of PCR Amplified Products Within Each Stock of *Catla catla* Generated From Four Decamer Primers of Random Sequences

Stocks	No. of Primers	Total No. of Amplified Products	Average No. of Bands Amplified per Primer	Total No. of Common Bands	Total No. of Polymorphic Bands	Percentage of Polymorphism (%)
Puri	4	42	10.50	24	18	42.85
Ganjam	4	36	9.00	19	17	47.22
Total	8	78	9.75	43	35	44.87
Average value	4	39	19.5	21.5	17.5	44.871

Table 10.19 Summary of the Results of One-Way ANOVA to Test for Difference in Intraspecies GS Values Among the Individuals of Each Species of *Catla* Calculated Based on RAPD Banding Patterns

Source of Variation	Degrees of Freedom	Sum of Squares	Mean Squares	F (Calculated)	Significance
Between species	34	0.115	0.003	0.432	0.966*
Error (within species)	10	0.078	0.008		
Total	44	0.194	0.011		

*Significant at $p < 0.001$.

the NTSYS pc 2.2 software with DICE coefficient. The interspecies GS value for Puri/Ganjam obtained through the program were estimated to be (all values per 1.000) 0.943 (Table 10.8 and Fig. 10.16). From the analyzed data it can be seen that both stocks are more genetically similar. with the highest GS value of 0.943.

10.3.6 Potential DNA Markers

The scorable DNA bands of molecular size 1100 and 1200 bp with primer OPC-02, 750 bp with OPC-16, 920 bp with OPA-18, and 1200 bp with OPA-11 were obtained having potential as a DNA marker for *C. catla* identification at molecular level of the two different stocks.

10.3.7 Cluster Analysis

UPGMA clustering method was used to generate a dendrogram for the two stocks of *C. catla* in Orissa in the present study by computing the GS values with DICE coefficient in NTSYSpc 2.2 program. The dendrogram showed one cluster with the Puri and Ganjam stocks.

Then the GS matrices for intraspecies variation were calculated with DICE coefficient with NTSYS software by pair-wise comparison of 10 individuals in each stock following the method of Nei and Li (1979). The GS values for each stock ranged from 0.67 ± 0.663 in Puri and 0.70 ± 0.599 in the case of Ganjam stock (Table 10.15). The

Table 10.20 Size of Each Amplified DNA Band in the Individuals of *Catla catla* With Primer OPA-11 of Puri District

Lanes / Rows	Lane 1 (mol.wt.)	Lane 2 (mol.wt.)	Lane 3 (mol.wt.)	Lane 4 (mol.wt.)	Lane 5 (mol.wt.)	Lane 6 (mol.wt.)	Lane 7 (mol.wt.)	Lane 8 (mol.wt.)	Lane 9 (mol.wt.)	Lane 10 (mol.wt.)	Lane 11 (mol.wt.)
Row 1	–	1400	1400	1400	1400	1400	1400	1400	1400	1400	1400
Row 2	–	1200	1200	1200	1200	1200	1200	1200	1200	1200	1200
Row 3	–	1100	1100	1100	1100	1100	1100	1100	1100	1100	1100
Row 4	1000	1000	1000	1000	1000	1000	1000	1000	1000	1000	1000
Row 5	900	900	900	900	900	900	900	900	900	900	900
Row 6	800	800	800	800	800	800	800	800	800	800	800
Row 7	700	–	–	–	–	–	–	–	–	–	–
Row 8	–	620	620	620	620	620	620	620	620	620	620
Row 9	600	–	–	–	–	–	–	–	–	–	–
Row 10	–	550	550	550	550	550	550	550	550	550	550
Row 11	500	500	500	500	500	500	500	500	500	500	500
Row 12	400	400	400	400	400	400	400	400	400	400	400
Row 13	–	–	–	320	–	–	–	320	320	–	–
Row 14	300	–	–	–	–	–	–	300	300	–	–
Row 15	–	280	280	–	280	280	280	–	–	280	280
Row 16	200	–	–	–	–	–	–	–	–	–	–
Row 17	100	–	–	–	–	–	–	–	–	–	–

Molecular weight standard: 100 bp ladder. Molecular weight unit: bp.

mean intraspecies GS values were 0.7897 ± 0.663 for Puri stock and 0.836 ± 0.599 in the case of Ganjam stock (Table 10.20). The highest GS values within the stock were obtained for Ganjam followed by Puri. These intraspecies GS values estimated for two stocks were checked by one-way ANOVA (SPSS version 16) and found to be significantly different at $p < 0.001$ (Table 10.2). Similarly the two representative DNA samples from each of the two stocks of catla were selected primer-wise for interspecies genetic variability analysis (Figs. 10.12–10.15). The matching coefficients of GS for the same four primers that amplified 78 bands were calculated using the NTSYS pc 2.2 software with DICE coefficient. The interspecies GS values obtained for Puri/Ganjam through the program were estimated to be (all values per 1.000) 0.943.

References

Carvalho, G.R., 1993. Evolutionary aspects of fish distribution: genetic variability and adaptation. J. Fish. Biol. 43 (Suppl. A), 53–73.

FAO Yearbook of Fishery Statistics, 1997. Vol. 81.

FAO Yearbook of Fishery Statistics, 2007. Vol. 101.

Jhingran, A.G., 1986. Artificial recruitment and fisheries management of Indian reservoirs. Bull. No. 45, CICFRI, Barackpore. 59 p.

Nei, M., Li, W.H., 1979. Mathematical models for studying genetic variation in terms of restriction endonucleases. Proc. Nat. Acad. Sci. U S A 74 (52), 67–5273.

Schiverwater, B., Str eit, B., Wagner, G.P., Desalle, R., 1997. Molecular Ecology and Evolution: Approaches and Applications. Birkhauser Verlag, Basal, Boston.495–508.

Singh, B.N., Das, R.C., Sahu, A.K., Kanungo, G., Sarkar, M., Sahoo, G.C., Nayak, P.K., Pandey, A.K., 2000. Balanced diet for broodstocks of Catla catla and Labeo rohita and induced breeding using ovaprim. Journal of Advanced Zoology, 21, 92–97.

Welsh, J., McClelland, M., 1990. Fingerprinting genomes using PCR with arbitrary primers. Nucl. Acids Res. 18, 7213–7218.

Williams, J.G.K., Kubelik, A.R., Livak, K.J., Rafalski, J.A., Tingey, S.V., 1990. DNA polymorphisms amplified by arbitrary primers are useful as genetic markers. Nucl. Acids Res. 18, 6531–6535.

Williams, J.K.G., Kubelik, A.R., Levak, K.J., Rafalsky, J.A., Tyngey, S.V., 1991. DNA polymorphism amplified by arbitrary primers are useful as genetic markers. Nucl. Acids Res. 18, 6531–6535.

PART

III

INNOVATIONS

APPROACHES TO GENETIC IMPROVEMENT

Aquatic animals allow the implementation of several approaches for genetic improvement. These include hybridization and cross-breeding, chromosome manipulation, sex control, transgenesis, and selective breeding. These are almost always mentioned in reports of aquaculture genetics, papers, and proceedings without making any further judgment about their relative practical value. For instance, it is seldom, if ever, emphasized that of all the genetic approaches only selective breeding offers the opportunity of continued genetic gain, that the gains accrued can be permanent. This again consolidates that it is the only approach through which not only the genetic gains can be transmitted from generation to generation, but at the same time gains in a nucleus can be multiplied and expressed in thousands or millions of individuals in the production sector (Ponzoni et al., 2007, 2008). So we may interpret that gains through selective breeding are more advantageous compared to other approaches that result in **"one off"** expressions of benefit. However, this may be applied at the multiplication (hatchery) level, but not at the nucleus level.

11.1 Different Approaches of Selection Program

11.1.1 General

It has already been established that for proper implementation of individual selection approaches following an increasing order of

Induced Fish Breeding. DOI: http://dx.doi.org/10.1016/B978-0-12-801774-6.00011-0

complexity, it is advisable to start with the simplest one. As expected, it is very much reasonable that the implementer generally proceeds with specific requirements that may constitute a limitation for their implementation, especially in developing countries. Along with this, it may happen that the traits of interest are endowed with genetic variation and also haven't suffered from any negative consequences like inbreeding, bottleneck, genetic drift, in which case the program will be at its least effective (Teichert-Coddington and Smitherman, 1988; Huang and Liao, 1990). This phenomenon could be considered repetitive and unnecessary in a livestock or crops context, but not in aquaculture where the application of quantitative genetics lags decades behind the two former fields.

11.1.2 Individual or Mass Selection

The terms "individual selection" and "mass selection" are often used interchangeably, and they refer to selection solely based on the individual's phenotype. It has now become a common strategy with fish because of its simplicity. It does not require individual identification or the maintenance of pedigree records, hence it may be considered the least costly method. In principle, it can produce rapid improvement if the heritability of the trait(s) under selection is high. Under those circumstances, however, there is risk of inbreeding due to inadvertent selection of progeny from a few parents producing the best offspring, especially if the progeny groups are large. For growth rate and morphological traits (easily assessed, expressed in both sexes) it can be quite suitable. By contrast, individual selection is not suitable for situations in which the estimation of breeding values requires slaughter of the animals (e.g., carcass and flesh quality traits) or challenge of some sort (e.g., selection for salinity tolerance or for disease resistance).

Hulata et al. (1986) carried out two generations of mass selection for growth rate with Nile Tilapia (*Oreochromis niloticus*) and observed no improvement over the original base population. They attributed the lack of response to selection to a number of possible factors, including inbreeding and genetic drift. They concluded that mass selection was not a promising method unless measures could be taken to control inbreeding. World Fish (unpublished) records indicate that the experience with Silver Barb (*Barbonymus gonionotus*) in Bangladesh and Thailand and Common Carp (*Cyprinus carpio*) in Vietnam has been of satisfactory response to selection in early generations up to the fourth or fifth, declining sharply thereafter.

Overall, the evidence of various experimental work suggests that simple, unstructured, mass selection will not register positive results unless the number of parents (founder population) is large (Gjerde et al., 1996; Villanueva et al., 1996), and even so, chance could have a negative effect. Some form of structuring to control the parental contribution to the next generation appears necessary. If controlled pair matings can be carried out, the results of Bentsen and Olesen (2002) can be used to formulate the design of the breeding program. Scientists in some of the Asian countries investigated the effect of number of parents selected and of number of progeny tested per pair for a definite range of population sizes and heritability values. The results indicate that rate of inbreeding can be maneuvered at 1% per generation when a minimum of 50 pairs is mated and the number of progeny tested from each pair is standardized to 30–50 progeny. The experimental results, barring identification of individual phenotype, elucidate the conduct of pair matings, primary maintenance of the progeny out of such matings in separate tanks, as well as controlled contribution of each full sibfamily to the next generation when fish are assigned to community rearing. As it is envisaged that, in some developing countries, implementation of such practices is not possible, efforts were initiated to develop suitable befitting technology.

11.1.3 Selection Within Cohorts and Exchange of Breeders

As reported by Eknath (1991), the primary cause for successive genetic deterioration in hatcheries in India is due to poor management of brood stock as well as the poor understanding about the scientific basis and principle of Induced breeding technology. It is suggested that available brood stock could be arbitrarily divided into several groups and then rotational matings between the individuals of different groups could be initiated to avoid inbreeding. With this understanding, scientists (Ponzoni et al., 2007, 2008) at world fish center, in an attempt at improvement, followed the mating design used by McPhee et al. (2004) for weight selection in red claw crayfish (*Cheraxquadri carinatus*). At the start the entire population was divided into cohorts, namely, groups sampled from a previously established foundation population. A cohort is a group of a population who share a common characteristic over a certain period of time or groups whose members share a significant experience at a certain period of time or have one or more similar characteristics. People born in the same year, for example, are the birth cohorts (generation) for that year. Similarly, married men or those who smoke are cohorts of the other married men or other smokers. After that a selection line was created, consisting of 20 cohorts, each having 15 females and 10 males from the foundation parents. A control line consisting of eight cohorts of the same size was also maintained. Generally, one hundred individuals were measured per cohort while offspring of individual cohorts were allowed to hatch and grow in separate pens in an experimental pond. At harvest, individuals which registered maximum weight gain from each cohort were chosen as parents of the next generation as a part of selection line, whereas individuals bearing average weight were chosen for the control line. In either case, selection was based on the difference between the harvest weight of an individual and its cohort mean. This within-cohort selection aims to eliminate the environmental effect of cohorts on growth differences among individuals. The same number of individuals was selected from each cohort. Animals selected in one cohort were mated with those selected in another one to avoid mating among related animals and inbreeding as well. After four generations of selection harvested weight registered 1.25 times greater weight gain compared to control.

As we know this type mating scheme does not unveil the exact number of parents contributing to the next generation and also the rate of inbreeding can be calculated only for the worst case, i.e., when only one pair per cohort contribute to the production of offspring. These findings let us understand that even if only one pair from each cohort produced progeny the inbreeding rate was not excessive, and thus it would be able to ensure that it would not create any problems due to inbreeding. With regards to the exchange of breeders between cohorts, this could be achieved by shifting the males born in one cohort to another in a pattern as described by Nomura and Yonezawa (1996), following, for instance, Cockerham's cyclical mating system (Cockerham, 1970). In practice, we have found that, in contrast to single pair matings, selection within cohorts with exchange of breeders between cohorts following a prescribed pattern is a feasible design even with limited resources. Field personnel feel comfortable with it, and will thus rigorously adhere to the instructions provided.

11.1.4 Methods of Selection for More Than One Trait

There are three methods of selecting for more than one trait: tandem selection, independent culling levels, and index selection.

11.1.4.1 Tandem Selection

This is selection for one trait or character at a time until it reaches an acceptable level followed by selection for a second trait, then a third trait, and so on. For instance, the milk yield of goats may be improved in the first case and then growth (meat production) would be addressed. Under tandem selection, if there is positive correlation between the traits to be considered, improvement can be realized in the second trait even as selection is applied only for the first trait. The disadvantage of this system is if a negative correlation exists between the two traits. In that case, performance of the second trait will decline as a result of selection for the first trait or selection for the second trait will erode progress made in the first trait.

11.1.4.2 Independent Culling Level

Selection of sheep and goats based on an independent culling level sets a certain accepted level of means for automatic culling of animals. It is like an examination system with different pass marks for each subject, but if the student fails in one subject, then he/she fails in all. There is no compensation for poor performance in one trait by superior performance in another. This method is most useful when there are a small number of traits (usually two) and where selection is done at different stages in an animal's life. For instance, we may cull some animals for poor performance in weaning weight and then later for reproductive performance. The disadvantage of this method is that exceptionally superior animals for one trait cannot be selected if they perform below the standard set for the second trait.

11.1.4.3 Index Selection

In an index selection, traits are combined to provide a single criterion merit, often economic-based. This type of selection is usually closer to the desire of farmers. With selection done on an index, deficiencies in any one trait can be compensated by outstanding performance in other traits an option which is not available when using independent culling levels. While index is the most efficient of the three methods, an index is the most complicated to create and requires a team of experts to construct the index weights.

References

Bentsen, H.B., Olesen, I., 2002. Aquaculture 204, 349.

Cockerham, C.C., 1970. In: Kojima, K. (Ed.), Mathematical Topics in Population Genetics Springer Verlag, New York, pp. 104.

Eknath, A.E., 1991. NAGA. ICLARM Q. 738, 13.

Gjerde, B., Gjoen, H.M., Villanueva, B., 1996. Livest. Prod. Sci. 47, 59.

Huang, C.M., Liao, I.C., 1990. Aquaculture 85, 199.

Hulata, G., Wohlfarth, G.W., Halevy, A., 1986. Aquaculture 57, 177.

McPhee, C.P., Jones, C.M., Shanks, S.A., 2004. Aquaculture 237, 131.

Nomura, T., Yonezawa, K., 1996. A comparison of four systems of group mating for avoiding inbreeding. Genet. Sel. Evol. 28, 141–159.

Ponzoni, R.W., Nguyen, H.N., Khaw, H.L., 2007. Aquaculture 269, 187.

Ponzoni, R.W., Nguyen, H.N., Khaw, H.L., Ninh, N.H., 2008. Aquaculture 285, 47.

Teichert-Coddington, D.R., Smitherman, R.O., 1988. Trans. Am. Fish. Soc. 117, 297.

Villanueva, B., Woolliams, J., Gjerde, B., 1996. Optimum designs for breeding programs under mass selection with an application to fish breeding. Anim. Prod. 63, 563–576.

CONSERVATION HATCHERY AND SUPPLEMENTATION—A RECENT APPROACH TO SUSTAINABLE AQUACULTURE

Conservationists opined that rearing any living creature in captivity and then subsequent release of the same into its natural habitat is a difficult process and involves application and integration of a number of rearing strategies, all of which are known individually to affect the inherent fitness of the creature to survive and breed in its natural ecosystem. For an aquatic creature, like fish, the conservation process is even more complex, as some strategies have to be executed within an invisible habitat and not under the usual captivity controls. A conservation hatchery may be defined as a rearing facility to breed and propagate a stock of fish with equivalent genetic resources of the native stock, and with the full ability to return to reproduce naturally in its native habitat. A conservation hatchery is therefore a facility equipped with a full complement of culture strategies to produce very specific stocks of fish in meaningful numbers. It may also permute particular strategies to match the particular requirements and biodiversity of any individual stock to its ecosystem and in this process while one combination of strategies helps restore a depressed stock, another helps reduce the risks of a certain supplementation program. The operation and management of conservation hatchery management should, therefore, be unique in time, stock-specific, and native to its habitat. Though true conservation hatcheries are not in existence till date, some production hatcheries are adopting some individual conservation strategies in an attempt to improve fitness and increase stock survival, but there is

Induced Fish Breeding. DOI: http://dx.doi.org/10.1016/B978-0-12-801774-6.00012-2

currently no single hatchery capable of applying a full package of strategies to produce a fish with the equivalent genetic resources of a local native stock. One reason is that the complete framework required for establishing a conservation hatchery has never been conceptualized. It presents the full complement of culture strategies available to hatchery managers. It recommends practical expedients to save depleted stocks, reform traditional hatcheries, and produce more adaptable juveniles to maximize the benefits and reduce the risks of supplementation programs. The principle of the primary framework for establishing a conservation hatchery was proposed first by a group of scientists of the National Marine Fisheries Services (NMFS), Northwest Fisheries Science Center (NWSFC), Resource Enhancement and Utilization Technology Division. Principally, it formulated guidelines for the management and operation procedure initially for establishing conservation hatcheries for Pacific salmon listed under the Endangered Species Act (ESA). Although initially the technique was in vogue for ESA-listed species of Columbia River Basin, it is equally applicable to any endangered stock in any ecosystem or habitat.

12.1 History of the Conservation Hatchery Concept

Modern production hatcheries are so instrumental in supplying fish seed to the common property resource that we can't think of separating culture management from that of hatchery management. Over 5 billion hatchery-reared juveniles are released annually into the Pacific Ocean from North American hatcheries. On the Columbia River alone, nearly 100 hatcheries produce about 200 million fish, which provide up to 80% of the fish in the leading fisheries. However, despite the great success of production hatcheries, the final decades of the 20th century saw the emergence of a different philosophy behind salmon resource management. Over 100 years ago, the 19th century was ending on an optimistic note for the recovery of depleted commercial fisheries stocks on both sides of the Atlantic Ocean. The new field of fisheries science had just been born out of scientific and public concern, and the new technology of artificial hatching was developing into the tool to make recovery possible. Today, after five generations of scientific endeavor and developing hatchery technology into an efficient farming process the 20th century ended in controversy and pessimism for the future of fisheries. Except for the years when wars prevented most commercial fishing, both inland and coastal fisheries have continued their decline unabated. Moreover, the decline is now on a global scale, with some stocks thought to be near the point of extinction; and hatchery technology has gone from panacea to patsy—the convenient inhuman scapegoat to shoulder the blame for human failure. But hatchery technology is not the only culprit. One hundred years ago, most of the early fisheries scientists recognized there was no single panacea for stock recovery (Wood, 1953). In 1901, e.g., the Royal Commission on Salmon Fisheries in England reviewed the issue of artificial hatching in Great Britain and North America, and its report in 1902 did not recommend artificial hatching (Calderwood, 1931). As one contributor at the hearings observed, "There is no example of the establishment or maintenance of a commercial salmon fishery upon any river in North America which has depended for its yield upon artificial culture, unsupported by restrictions upon netting or by accessible spawning grounds." Clearly, these pioneers recognized that fisheries management and fish habitat are equally important prerequisites in the game of stock recovery. Artificial culture, however, proved to be a powerful piece on the board and one easy to overplay. Moreover, the intricacies, boundaries, and length of the game were never fully understood; and some of the rules had only started to emerge in the final decades of the 20th century. With population and economic pressures tightening around the environment after World War II, the existence of many species became increasingly put at risk on a number of fronts all at the same time. Firstly, for every aquatic species the principal threat was the increasing and irreversible loss of habitat. This was due to either industrial and social progress, resulting in physical degradation and pollution, or from well-intentioned biological interventions creating an imbalance in the endemic populations in the ecosystem. Secondly, for aquatic species with any commercial value, an added threat was annual overharvesting. With more well-intentioned attempts to control overharvesting, and in compliance with regional fishing agreements promulgated by international organizations and fisheries commissions, increasing emphasis for stock recovery was being put on fisheries management. However, yet another risk for all species in this milieu became unwise fisheries management practices. The response to the growing risks to all ecosystems from these threats and other types of human interventions was the ESA. The 1973 Act recognized that, "various species of fish, wildlife, and plants in the United States have been rendered extinct as a consequence of economic growth untempered by adequate concern and conservation." For Pacific salmon the emphasis of the ESA was on habitat conservation and the protection and recovery of the natural populations. But if habitat restoration was a critical part of any plan for conservation and recovery, it was no easy task. Firstly, the aquatic ecosystem for any diadromous fish was multidimensional and linked inseparably to a surrounding terrestrial ecosystem. Secondly, the biodiversity of each natural population was an absolute indispensable necessity for its survival. Therefore, if species biodiversity was to be maintained then genetic integrity of the species must be conserved and, in some cases, quickly. Consequently, the ESA acknowledged that conservation and recovery might still depend on artificial culture practices. Several of the roles and actions of artificial propagation in recovery plans were later summarized by Hard et al. (1992). The actions included the choice of donor stock, brood stock collection and mating, husbandry techniques, release strategies, monitoring and evaluation, and captive brood stock programs. Since the ESA was enacted, maintaining the biodiversity and genetic integrity of species of fish when the commercial fishery is enhanced by artificial hatching has proved to be a very complex and controversial issue, and one that has stretched the minds of many multidisciplinary groups of scientists. In the mid-1980s, the National Research Council (NRC) of the United States National Academy of Science formed a Committee on Managing Global Genetic Resources and established an Aquatic Animal Working Group. Perhaps symptomatic of the complexity of the issues the Group could not produce a report for the Committee and the information was not published until 1995 (Thorpe et al., 1995). Although the Group included Atlantic salmon as one of four species reviewed at length, the six principal recommendations were general for effective management and conservation of all commercial aquatic species. No specific recommendations were addressed to the role of artificial culture, but the Group noted there was great scope for improvement in hatchery management, "especially in conservation hatcheries, to ensure that the initial genetic diversity was comparable to that in the wild stock to be augmented, and that the hatchery products were ecologically competent." The Group also recognized that hatcheries might be the only tractable solution to the conservation and recovery of many stocks. In 1992 the NRC tried again. This time, through the Committee on Protection and Management of Pacific Northwest Anadromous Salmonids, the mandate was more specific and there was a full and scientific evaluation of hatcheries and their roles. At about the time the study ended, and the report of the Committee was being prepared in book form entitled, *Upstream* (NRC, 1996), two other scientific perspectives of artificial hatching were being undertaken. One was the National Fish Hatchery Review Panel, which published its findings in 1994 (NFHRP, 1994), and the other was an Independent Scientific Group commissioned by the Northwest Power Planning Council (NWPPC). The report of this Group (ISG, 1996) was published as a book called *Return to the River*.

12.2 Conclusion and Recommendation

The different scientific groups were almost identical, and in a review of salmonid artificial production in the Columbia River Basin, commissioned by the NWPPC, their findings were evaluated and compared by a Scientific Review Team. In its report (SRT, 1998), the team identified the following 10 points as the common denominators of these documents: (1) Hatcheries have generally failed to meet their objectives. (2) Hatcheries have imparted adverse effects on natural populations. (3) Managers have failed to evaluate hatchery programs. (4) Hatchery production was based on untested assumptions. (5) Supplementation should be linked with habitat improvements. (6) Genetic considerations have to be included in hatchery programs. (7) More research and experimental approaches are required. (8) Stock transfers and introductions of nonnative species should be discontinued. (9) Artificial production should have a new role in fisheries management. (10) Hatcheries should be used as temporary refuges, rather than for long-term production. The team recommended that the first seven of these elements should always be considered in the development of hatchery policies, and it went on to make 21 individual recommendations for production technology practices, hatchery management, and research and monitoring. The issues regarding artificial hatching were not the sole prerogative of the large scientific groups commissioned to study the problems. For the last 20 years individual or groups of issues have been the subject of independent peer-reviewed papers written mostly by fisheries biologists and geneticists. Dentler and Buchanan (1986) proposed a comprehensive review of the role of salmonid hatcheries and their relationship to wild salmon stocks. They concluded that hatchery production was necessary in many areas and hatcheries would continue to be a useful tool of fishery managers in the future, but they believed a more cautious and critical examination might reveal better ways to integrate wild and hatchery production with associated fisheries. Brannon (1993) subsequently stated in a critical analysis of what he called the perpetual oversight of hatchery programs that by neglecting the requirements of natural populations, and consequently their adaptive environmental traits, hatchery programs produced fish which had little chance of integrating into the ecosystem. Reisenbichler and Rubin (1999) concluded that the only similarities in hatchery and wild environments for salmon were water and photoperiod. Everything else, such as food, substrate, density, temperature, flow regime, competitors, and predators, was dissimilar. In an analysis of salmon and steelhead supplementation, Miller (1990) studied over 300 projects designed to supplement these fisheries in North America. Among the many observations he commented that success at rebuilding runs was scarce. Projects were more successful at just returning fish. Moreover, adverse impacts to wild stocks had been shown or postulated from about every type of hatchery introduction, even though the goal of the project was to rebuild a run. Overstocking with hatchery fish might be one of the most significant problems in supplementation projects. He concluded that there were no guarantees that hatchery supplementation could replace or consistently augment natural production. Always the top priority was the protection and nurturing of the natural runs. The issue of supplementation of wild salmon stocks was taken up by Sterne (1995). He examined the scientific, legal, and policy positions of those groups for and against the practices of supplementation in the Columbia River Basin. He identified a number of common areas for agreement, which included both supplementation and nonhatchery alternatives, such as a moratorium on ocean and in-stream fishing. But he concluded that habitat restoration and protection were the real cornerstones of supplementation. The primary goal of any program in the Basin was the biological health of the stocks, and therefore supplementation should take place only under carefully controlled conditions at nonpermanent facilities in an adaptive, experimental framework.

12.3 Recovery of Threatened and Endangered Species

In a paper on conservation aquaculture and endangered species, Anders (1998) discussed some basic tenets of endangered species management from the perspective of population biology, the role of conservation aquaculture in endangered species management, and the potential dangers of fisheries management policies which consider aquaculture as a last resort for conserving endangered fish populations. He defined conservation aquaculture as the use of aquaculture for conservation and recovery of endangered fish populations. It did not embrace standard hatchery practices and tried to eliminate as much artificial conditioning as possible. It contrasted directly with the ideology underlying more-traditional hatchery supplementation programs, which measure success by the numbers of fish released from a hatchery. He believed that the use of aquaculture for conservation and recovery should be based on objective science, not risk anxiety. More recently, Waples (1999) identified and analyzed several myths and misconceptions about hatcheries and their effects on natural populations, which were impeding progress and contributing to the controversy about hatcheries. He concluded that hatcheries were intrinsically neither good nor bad, and their value could be determined only in the context of clearly defined goals. He believed that genetic changes in cultured populations could be reduced but not eliminated entirely. Some risks from hatcheries were overstated, but there was some empirical evidence that hatcheries had many adverse effects. The priority was for risk-averse hatchery programs, supported by effective monitoring and evaluation. In conclusion, he recommended four areas where actions could resolve some of the controversies regarding fish hatcheries. These were identifying goals, conducting overall cost–benefit analyses to guide policy decisions, improving the information base, and dealing with uncertainty. Backed by the increasing scientific concerns over traditional hatchery practices, the shift toward a conservation ethic has increased. The theme has become evident in all the new plans being formulated by involved agencies and management groups. The proposed recovery plan for Snake River salmon (Schmitten et al., 1995) opened with the statement that the focus of the plan differed from previous management strategies. New emphasis was being placed on natural fish escapement, improved migration conditions for juveniles and adults, increased riparian area protection, and equitable consideration of natural fish in resource allocation processes. The goal of the plan was now to restore the health of the Columbia and Snake River ecosystem and to recover listed Snake River salmon stocks. Implementation of the plan would conserve biodiversity, a factor which was essential to ecosystem integrity and stability. With specific regard to artificial propagation, the plan called for ecological interactions between hatchery fish and wild fish to be minimized but to use culture technologies to conserve remaining gene pools through captive brood stock, supplementation, and gene bank programs. Similarly, to meet the objectives of the artificial production and evaluation plan for summer chum populations in the Hood Canal and Strait of Juan de Fuca regions, the tri-agency Supplementation Work Group (SWG, 1998) produced a set of general principles calling for maintenance of natural population characteristics for fish taken into the hatchery environment. In general, these principles were previously identified by Kapuscinski and Miller (1993), and included procedures for brood stock collection and spawning, incubation, juvenile rearing, smolt release, and monitoring and evaluation. The plan also contained a set of specific criteria guiding supplementation and reintroduction program operations, which included donor-stock selection and collection methods, spawning and mating protocols, incubation and rearing protocols, and release strategies. At the same time, the conservation theme was also becoming more evident in policy documents. The NWPPC, in preparation for developing its policy recommendations for the US Congress, drafted a policy statement for artificial production by the Columbia River Basin Hatcheries (NWPPC, 1999), which recognized that artificial production was undergoing a major transition. This conclusion was based on the findings of the Artificial Production Review, an initiative the Council had taken earlier to facilitate further discussion and assist in developing a coordinated policy for the hatcheries in the Basin. Program managers, the review team said, were under pressure to transform hatcheries to widen the harvest opportunities, reduce the adverse impact of hatchery production on wild fish, and attempt to use artificial production techniques to try to rebuild naturally sustaining populations. The draft statement reinforced the Council's objective to develop policies consistent with eight previously identified scientific principles, one of which stated that the abundance and productivity of fish and wildlife should reflect the conditions they experience in their ecosystems over the course of their lifecycle. Furthermore, two basic premises for its initiatives regarding artificial production included mimicking wild population rearing conditions to improve survival, and mimicking natural rearing conditions to reduce impacts on wild populations. The draft went on to outline a series of policies for artificial production performance standards, ecological interactions, genetics, and fish health, and concluded with a section on the use of the tools in planning. From these and other converging events and observations, it is clear that the pressing need to preserve indigenous wild salmon stocks has brought in a new era of conservation. Many stocks are now protected (ESA-listed) within their native habitats, and restrictions imposed by ESA mandates have a major impact on the management and operation of production hatcheries, and traditional users of hatchery fish. A conservation hatchery can play a vital part in the recovery of threatened and endangered species by maintaining their genetic diversity and natural behavior, and reducing the short-term risk of extinction. The use of conservation measures in the management of salmonid fisheries is not a new idea. Several measures have already been proposed by fisheries managers in the Pacific Northwest. These include, e.g.:

- Enactment of strict laws prohibiting unauthorized nonindigenous fish stock transfers (intentional transplantation);
- Regular review of health, genetic, and smolt quality problems and management;
- Mass mark hatchery fish for identification;
- Application of effective reintroduction to save native gene pool;
- Application of, if necessary, production caps to match release numbers with the finite carrying capacity of fresh and saltwater habitats;
- Captive brood stock development through selective breeding techniques should be initiated to avoid the risk of extinction and to conserve endangered gene pools;
- Procedure to reduce selection for domestication by introducing more natural rearing protocols;

- Produce better-quality smolts indistinguishable from their wild counterparts;
- Development/modification/refinement of hatchery techniques with an aim to reduce harmful postrelease interactions between wild and hatchery fish.

NWSFC spent over a decade conducting a definite goal-oriented research program involving the said criteria. Guiding much of the Center's research is the fundamental hypothesis that successful stock recovery depends on the quality of juveniles reared in the hatchery. Consequently, the underlying objectives of its many research projects in all aspects of artificial propagation are to identify and maintain the attributes of wild fish. Artificial rearing conditions within a hatchery, it is now recognized, can produce fish distinctly different from wild cohorts in behavior, morphology, and physiology. Hatchery methodologies can impose different selective pressures on fish and these can change overall fitness in many ways. Conventional hatchery rearing practices can alter genetic fitness through spawning and fertilization protocols. Hatcheries can inadvertently select for fish adaptable to high densities and feeding levels, and fish which cannot adapt may be selected against and not survive to release. Similarly, conventional practices purposely reduce individual size variability. Within a hatchery population this may be desirable, but in the long term this can be detrimental if fish are expected subsequently to rear and spawn in the wild. The wide natural variability in development and timing characteristic of wild fish may be inherent factors which enable them to adapt to changing freshwater and marine conditions. With the emphasis on wild fish required under the ESA, there is an opportunity to transfer the role of certain hatcheries from production to conservation. A conservation hatchery will operate on the concept that high-quality fish, behaviorally and physiologically similar to wild cohorts, can be produced in conditions which simulate the natural life histories of each particular species under culture. Scientific information now available makes it feasible and practical for a hatchery to propagate juveniles similar in growth, development, and behavior to their wild cohorts. The probability of success is high. Animal behaviorists have shown that behavioral repertoires can often be recovered even after many generations, simply by providing appropriate environmental stimuli. The following sections identify the major culture strategies for the management and operation of conservation hatcheries in the Pacific Northwest, and outline potential protocols. The strategies are based on a combination of modern conservation principles and basic salmonid biology. They are common sense, logical approaches to the needs for culturing wild-like animals. Some are backed by scientific research; others are currently being researched. However, any of these particular strategies can be implemented by

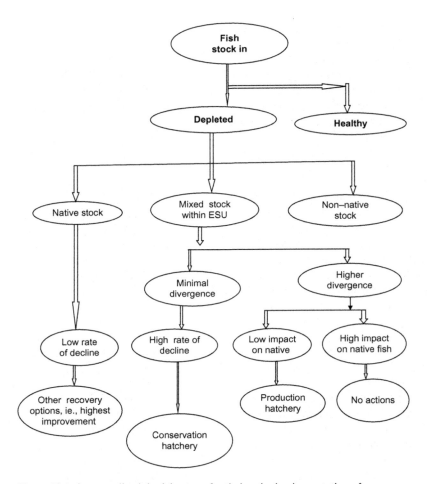

Figure 12.1 A generalized decision tree for timing the implementation of conservation hatchery.

hatchery managers using the latest and best scientific information. A generalized "decision tree" for timing the implementation of conservation hatchery strategies is illustrated in Fig. 12.1. The concerns include the status of the population, its genetic composition, rate of decline, and the impact of any actions on native fish. The needs and concerns of each conservation program will therefore be site-specific. They will also depend on the physical and management limitations of each individual hatchery. Consequently, the exact application of conservation hatchery strategies will depend on the particular stock of fish, its level of depletion, and the biodiversity of the ecosystem.

Genetic management guidelines, to maintain the genetic health of hatchery populations, though established but their implementation in fish hatcheries is limited or we may say that aquaculture activities as a whole, particularly of the fish breeding activities, are still untouched by the advanced emerging technologies. In some cases when applied, several negative genetic consequences are found to develop, mainly due to profit-making approaches, illiteracy on the part of farmers, as well as ignorance due to faulty dissemination from the initial stage. These include high levels of inbreeding, and outbreeding, loss of adaptation to captivity, reduced viability and fecundity, and reduced effective population size, depression in growth, reduced growth rate, genetic drift, introgression, that finally lead to the development of decreased fitness, loss of reproductive capability resulting in loss of population size and supplementation of wild populations (Ryman and Laikre, 1991; Araki et al., 2007a,b,c; Frankham, 2008). It is understood now that with an intensive genetic management program involving supplementation and conservation hatcheries, many of the negative genetic changes to wild fish populations may be mitigated. For example, the implementation of a pedigree-based breeding and genetic management program to maximize gene diversity and limit inbreeding at the same time, thus founding gene diversity of the captive population, can be preserved and at the same time help in maintaining the effective population size (Lacy, 1994; Ballou and Lacy, 1995). The winter-run Chinook salmon conservation hatchery provides a good example of a successful genetic management plan. By attempting to equalize founder contributions and ensuring that the hatchery did not produce a large fraction of the next generation, the supplementation of winter-run Chinook salmon into the wild population did not appear to decrease the overall wild effective population size (Hedrick and Hedgecock, 1994; Hedrick et al., 2000b). However, conservation hatchery populations have also been shown to accumulate negative genetic changes (Hedrick et al., 2000a,b; Osborne et al., 2006; Fraser and Bernatchez, 2008). Hedrick et al. (2000a) evaluated the bony tail chub captive brood stock and discovered low genetic diversity due to a small number of founders. In addition, analysis of the Rio Grande silvery minnow propagation program revealed that it maintained allelic diversity but still resulted in higher inbreeding in captive versus wild fish stocks, although a more recent study showed that the program has retained diversity in the captive and wild populations over the past decade (Osborne et al., 2006, 2012). These results demand a dire need for rigorous genetic management of captive populations to preserve their genetic integrity. This evaluation of the genetic management plan of the captive delta smelt population aims to assess its ability to minimize mk and minimize genetic divergence from the wild population, in an effort to inform conservation hatchery genetic management plans of other species.

References

Anders, P.J., 1998. Conservation aquaculture and endangered species. Fisheries 23 (11), 28–31.

Araki, H., Ardren, W.R., Olsen, E., Cooper, B., Blouin, M.S., 2007a. Reproductive success of captive-bred steelhead trout in the wild: evaluation of three hatchery programs in the hood river. Conservation Biology. 21, 181–190. [PubMed].

Araki, H., Cooper, B., Blouin, M.S., 2007b. Genetic effects of captive breeding cause a rapid, cumulative fitness decline in the wild. Science 318, 100–103. [PubMed].

Araki, H., Waples, R.S., Ardren, W.R., Cooper, B., Blouin, M.S., 2007c. Effective population size of steelhead trout: influence of variance in reproductive success, hatchery programs, and genetic compensation between life-history forms. Molecular Ecology 16, 953–966. [PubMed].

Ballou, J., Lacy, R.C., 1995. Identifying genetically important individuals for management of genetic diversity in pedigreed populations. In: Ballou, J., Gilpin, M., Foose, L. (Eds.), Population Management for Survival and Recovery: Analytical Methods and Strategies for Small Population Conservation. Columbia University Press, New York, NY, pp. 76–111.

Brannon, E.L., 1993. The perpetual oversight of hatchery programs. Fisheries Research, 18, 19–27. Elsevier BV. http://dx.doi.org/10.1016/0165-7836(93)90037-8.

Calderwood, W.L., 1931. Salmon hatching and salmon migrations The Buckland Lectures for 1930. Edward Arnold & Co., London, 95 p.

Dentler, J.L., Buchanan, D.V., 1986. Are Wild Salmonid Stocks Worth Conserving? Oregon Department of Fish and Wildlife Research and Development Section. Information Reports, No. 86-7. https://nrimp.dfw.state.or.us/CRL/Reports/Info/86-7.pdf.

Frankham, R., 2008. Genetic adaptation to captivity. Mol. Ecol. 17, 325–333.

Fraser, D.J., Bernatchez, L., 2008. Ecology, evolution and the conservation of lake-migratory brook trout: a perspective from pristine populations. Transactions of the American Fisheries Society in press.

Fraser, D.J., Bernatchez, L., 2008. Ecology, evolution and the conservation of lake-migratory brook trout: a perspective from pristine populations. Trans. Am. Fish. Soc. 137, 1192–1202.

Hard, J.J., Jones Jr., R.P., Delarm, M.R., Waples, R.S., 1992. Pacific Salmon and Artificial Propagation Under the Endangered Species Act. U.S. Department of Commerce, NOAA Tech. Memo., NMFS-NWFSC-2, 56 p.

Hedrick, P.W., Hedgecock, D., Hamelberg, S., Croci, S.J., 2000a. The impact of supplementation in winter- run Chinook salmon on effective population size. J. Hered. 91, 112–116.

Hedrick, P.W., Rashbrook, V.K., Hedgecock, D., 2000b. Effective population size in winter-run chinook salmon based on microsatellite analysis of returning spawners. Can. J. Fish. Aquatic Sci. 57, 2368–2373.

Hedrick, P.W., Hedgecock, D., 1994. Effective population size in winte r-run chinook salmon. Conserv. Biol. 8, 890–902.

The Independent Scientific Group, 1996. Return to the River: Restoration of Salmonid Fishes in the Columbia River Ecosystem. Prepublication copy: https://www.nwcouncil.org/media/7148863/96-6.pdf.

Kapuscinski, A.R., Miller, L.M., 1993. Genetic hatchery guidelines for the Yakima/Klickitat Fisheries Project. Co-Aqua, 2369 Bourne Avenue, St. Paul, MN.

Lacy, R.C., 1994. Managing genetic diversity in captive populations of animals. In: Bowles, M.L., Whelan, C.J. (Eds.), Restoration of endangered species. Cambridge University Press, Cambridge (UK).

Miller, W.H. (ed). 1990. Analysis of salmon and steelhead supplementation. Report to the Bonneville Power Administration, Contract DE-A179-88BP92663, 202 p. (Available from the Bonneville Power Administration, Box 3621, Portland, OR 97208).

NFHRP (National Fish Hatchery Review Panel), 1994. Report of the National Fish Hatchery Review Panel, 1994. The Conservation Fund, Arlington, VA.

NRC (National Research Council), 1996. Upstream: Salmon and Society in the Pacific Northwest. National Academy Press, Washington, DC.

NWPPC (Northwest Power Planning Council), 1999. Artificial Production Policy Statement for the Columbia Basin Hatcheries: A Program in Transition. February 17, 1999. Northwest Power Planning Council, Portland, OR, 15 p.

Osborne, M.J., Benavides, M.A., Aló, D., Turner, T.F., 2006. Genetic effects of hatchery propagation and rearing in the endangered Rio Grande Silvery Minnow. Rev. Fish Sci. 14, 127–138.

Osborne, M.J., Carson, E.W., Turner, T.F., 2012. Genetic monitoring and complex population dynamics: insights from a 12-year study of the Rio Grande silvery minnow. Evol. Appl. 5, 553–574. [PMC free article] [PubMed].

Reisenbichler, R.R., Rubin, S.P., 1999. Genetic changes from artificial propagation of Pacific salmon affect the productivity and viability of supplemented populations. ICES J. Mar. Sci. 56 (4), 459–466.

Ryman, N., Laikre, L., 1991. Effects of supportive breeding on the genetically effective population size. Conserv. Biol. 5, 325–329.

Schmitten, R.A., Stelle Jr., W., Jones, R.P., 1995. Proposed recovery plan for Snake River salmon. U.S. Department of Commerce, NOAA, Washington, D.C, 347 p.

Scientific Review Team, 1998. Review of salmonid artificial production in the Columbia River Basin. Report 98-33, Northwest Power Planning Council. Portland, OR., 77 p. (Available from Northwest Power Planning Council, 851 S.W. Sixth Avenue, Portland, OR 97204-1348).

Sterne, J.K., 1995. Supplementation of wild salmon stocks: a cure for the hatchery problem or more problem hatcheries? Coastal Manage. 23, 123–152.

SWG (Supplementation Work Group), 1998. Artificial Production and Evaluation Plan for Summer Chum Populations in the Hood Canal and Strait of Juan de Fuca Regions. US Fish and Wildlife Service, Washington Department of Fish and Wildlife, and Point No Point Treaty Council, Washington, DC.

Thorpe, J.E., Gall, G.A.E., Lannan, J.E., Nash, C.E., 1995. Conservation of fish and shellfish resources: managing diversity. Academic Press Limited, London, 206 p.

Waples, R.S., 1999. Dispelling some myths about hatcheries. Fisheries 24, 12–21.

Wood, E.M., 1953. A century of American fish culture, 1853–1953. Prog. Fish-Cult. 15 (4), 147–162.

INDEX

Note: Page numbers followed by "*b*," "*f*," and "*t*" refer to boxes, figures, and tables, respectively.

Printed in the United States
By Bookmasters